Two-Phase Flow and Heat Transfer

P. B. Whalley

Department of Engineering Science, University of Oxford

Series sponsor: **ZENECA**

ZENECA is a major international company active in four main areas of business: Pharmaceuticals, Agrochemicals and Seeds, Specialty Chemicals, and Biological Products.

ZENECA's skill and innovative ideas in organic chemistry and bioscience create products and services which improve the world's health, nutrition, environment, and quality of life.

ZENECA is committed to the support of education in chemistry and chemical engineering.

OXFORD NEW YORK TOKYO
OXFORD UNIVERSITY PRESS
1996

Oxford University Press, Walton Street, Oxford OX2 6DP

Oxford   New York
Athens   Auckland   Bangkok   Bombay
Calcutta   Cape Town   Dar es Salaam   Delhi
Florence   Hong Kong   Istanbul   Karachi
Kuala Lumpur   Madras   Madrid   Melbourne
Mexico City   Nairobi   Paris   Singapore
Taipei   Tokyo   Toronto

and associated companies in
Berlin   Ibadan

Oxford is a trade mark of Oxford University Press

Published in the United States
by Oxford University Press Inc., New York

A catalogue record for this book is available from the British Library

Library of Congress Cataloging in Publication Data
(Data applied for)
ISBN   0 19 856444 9

Typeset by the author using LaTeX
Printed in Great Britain by
The Bath Press, Avon

Series Editor's Foreword

The Oxford Chemistry Primers are now well established as popular undergraduate texts. The Primers are designed to provide concise introductions to a wide range of topics that may be encountered by chemistry students, and contain only the essential material that would normally be covered in an 8–10 lecture course. The Primer Series has now been extended to include topics in chemical engineering, and this Primer represents the first title in this new subject area.

In the near future, we can expect the discipline of chemical engineering to undergo rapid changes in terms of both the content of undergraduate courses and the nature of chemical research. The forthcoming titles in this series will cover new topics which the modern-day chemical engineering student should be conversant with, as well as topics which undergraduates traditionally find difficult to understand. In this first Chemical Engineering Primer, Dr Peter Whalley tackles a subject which many students find challenging—that of two-phase flow and heat transfer. Dr Whalley gives a concise introduction to this topic with clear definitions of relevant symbols and terminology. The models used to describe a variety of two-phase flow and heat transfer phenomena are described and their limitations are discussed. This Primer provides the undergraduate with a concise overview of the important concepts in two-phase flow and heat transfer and, just as importantly, a working understanding of our limitations in measuring and modelling these phenomena.

Lynn F. Gladden
Department of Chemical Engineering, University of Cambridge

Preface

Much of the material in this present work appeared in an earlier book published by OUP in 1987 (P. B. Whalley: *Boiling, Condensation and Gas–Liquid Flow*). This book is now out of print, and so the opportunity has been taken to produce a revised and shortened version for the Primer Series. The material omitted was about specific flow patterns (particularly annular flow), critical two-phase flow, post-burnout heat transfer, rewetting, and industrial equipment.

Like the earlier book, this one is intended for final year undergraduates, or graduate students in mechanical and chemical engineering. The objective has been to produce a book which is a simple introduction to two-phase flow and heat transfer, and provides an outline of the calculation methods which are available for producing answers to common problems. References to source materials and to more specialized works are given at the end of this book.

Most practical heat transfer problems involve phase changes: single phase heat transfer has been well studied and there are reasonably well established methods for solving most problems. In contrast most two-phase problems are not well understood at all. Pressure gradients in two-phase flow cannot be calculated to an accuracy better than 40%, and calculated values of boiling and condensing heat transfer coefficients are even less accurate.

I should particularly thank those who have taught me most about these subjects: Geoff Hewitt and the late John Collier. To work with them both at Harwell in the 1970s was unforgettable. Finally I should thank my family for their patience during the writing of this book.

Oxford
December 1995

P. B. W.

Contents

1   Introduction

1.1   Two-phase flow and heat transfer

Two-phase flows are commonly found in industrial processes and in ordinary life. The phases may of course be solid, liquid, or gas.

Gas–liquid flow occurs in boiling and condensation operations, and inside many pipelines which nominally carry oil or gas alone, but which actually carry a mixture of oil and gas.

Liquid–liquid flow (i.e. flow of two immiscible liquids) occurs, for example, in liquid–liquid extraction processes.

Gas–solid flow occurs in a fluidized bed and in the pneumatic conveying of solid particles.

Solid–liquid flow occurs during the flow of suspensions such a river bed sediments and coal-water slurry.

This book is concerned exclusively with gas–liquid flow, although some of the principles and methods can be applied to other types of two-phase flow. The first part of the book deals with adiabatic flow, that is flow with no heat addition or removal. This generally means that the gas and liquid flow rates are constant, although in high speed flow (as in critical flow) partial vaporization of the liquid may occur even though there is no heat addition. The remainder of the book deals with heat transfer in two-phase situations: boiling and condensation. The objective here is to identify the various types of heat transfer and to show that the heat transfer rate can be calculated.

Types of two-phase flow

Gas–liquid flow is the main topic here

1.2   Units

Throughout the text the SI system of units is used. Therefore almost all the equations are dimensionally consistent and there is no need for the 'gravitational constant' g_c. On the rare occasions when dimensionally incorrect equations are used the units of the variables in the equation are clearly specified. Experimental results are also quoted in SI units, the only exception is that the 'bar' is often used as the unit of pressure when displaying results.

$1 \text{ bar} = 10^5 \text{N/m}^2$

1.3   Nomenclature

The symbols used are defined when they are introduced. However some of the more common symbols are those for flow variables:

Physical property variables:
- ρ density (kg/m^3)
- μ viscosity (Ns/m^2)
- κ thermal conductivity (W/mK)
- C_p specific heat (at constant pressure) (J/kg K)

Quality is also sometimes referred to as the mass dryness fraction.

subscripts:
- ℓ liquid phase
- g   gas (or vapour) phase

V is the superficial velocity, and u is the actual velocity.

G total mass flux (kg/m^{2}s) = total (liquid + gas) mass flow rate (kg/s) divided by total channel cross-sectional area (m^2);

G_ℓ liquid mass flux (kg/m^{2}s) = liquid mass flow rate (kg/s) divided by total channel cross-sectional area (m^2);

G_g gas mass flux (kg/m^{2}s) = gas mass flow rate (kg/s) divided by total channel cross-sectional area (m^2);

x quality—this is the fraction of the mass flow rate which is gas, and so

$$x = \frac{G_g}{G} \tag{1.1}$$

V_g superficial velocity of the gas (m/s)—this is the velocity if the gas in the two-phase flow was flowing along in single phase flow in the channel, so

$$V_g = \frac{G_g}{\rho_g} \tag{1.2}$$

V_ℓ superficial velocity of the liquid (m/s), similarly defined, so

$$V_\ell = \frac{G_\ell}{\rho_\ell} \tag{1.3}$$

α void fraction—this is the time-averaged fraction of the cross-sectional area or of the volume which is occupied by the gas phase;

u_g actual velocity of the gas phase (m/s)—this is the velocity which would be measured if the velocity of a small volume of gas could actually be determined, so

$$u_g = \frac{V_g}{\alpha} \tag{1.4}$$

u_ℓ actual velocity of the liquid phase (m/s), so

$$u_\ell = \frac{V_\ell}{1 - \alpha} \tag{1.5}$$

1.4   Scope of the text

The scope of this book is inevitably limited as it is intended to be an introductory text. Fur further information the reader is referred in particular to the following books.

1. Hetsroni (1982) for a detailed survey of almost all aspects of two-phase flow and heat transfer.

2. Collier and Thome (1994) for a good summary of the available information about convective boiling.

3. Wallis (1969) for a wide coverage of topics in adiabatic two-phase flow.

4. Hewitt and Hall Taylor (1971) for a detailed description of annular two-phase flow.

5. Hewitt (1978) for a description of instrumentation and flow measurement techniques for two-phase flow.

6. Whalley (1987) for an expanded version of this text, including more material about boiling and condensation, and details of individual flow patterns.

Later chapters in this book refer to specific parts of these books, other books, research papers, and reports.

2   Two-phase flow patterns and flow pattern maps

2.1   Introduction

In gas–liquid flow the two phases can adopt various geometric configurations: these are known as flow patterns or flow regimes. Important physical parameters in determining the flow pattern arc:

1. surface tension, which keeps channel walls always wet (unless they are heated when they are usually wet) and which tends to make small liquid drops and small gas bubbles spherical; and

2. gravity, which (in a non-vertical channel) tends to pull the liquid to the bottom of the channel.

The common flow patterns for vertical upflow, that is where both phases are flowing upwards, in a circular tube are illustrated in Fig. 2.1. As the quality is gradually increased from zero, the flow patterns obtained are:

Surface tension and gravity help to determine the flow pattern.

1. bubbly flow, in which the gas (or vapour) bubbles are of approximately uniform size;

2. plug flow (sometimes called slug flow), in which the gas flows as large bullet-shaped bubbles. (There are also some small gas bubbles distributed throughout the liquid.);

Vertical flow patterns are bubbly, plug, churn, and annular flow.

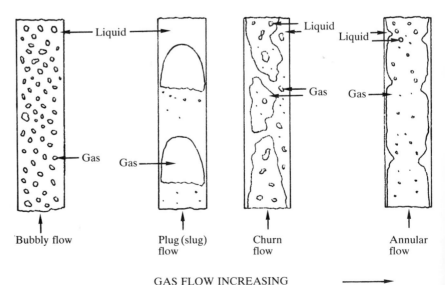

GAS FLOW INCREASING ⟶

Fig. 2.1 Flow patterns in vertical upflow in a tube.

3. churn flow, which is highly unstable flow of an oscillatory nature: the liquid near the tube wall continually pulses up and down;

4. annular flow, in which the liquid travels partly as an annular film on the walls of the tube and partly as small drops distributed in the gas which flows in the centre of the tube.

The common flow patterns for horizontal flow in a round tube are illustrated in Fig. 2.2. As the quality is gradually increased from zero, the flow patterns obtained are:

1. bubbly flow, in which the gas bubbles tend to flow along the top of the tube;

2. plug flow, in which the individual small gas bubbles have coalesced to produce long plugs;

3. stratified flow, in which the liquid–gas interface is smooth. Note that this flow pattern does not usually occur, the interface is almost always wavy as in wavy flow.

4. wavy flow, in which the wave amplitude increases as the gas velocity increases;

5. slug flow, in which the wave amplitude is so large that the wave touches the top of the tube; and

6. annular flow, which is similar to vertical annular flow except that the liquid film is much thicker at the bottom of the tube than at the top.

Many writers define other flow patterns, and nearly a hundred different names have been used. Many of these are merely alternative names,

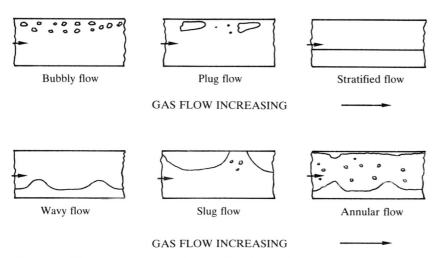

Bubbly flow   Plug flow   Stratified flow

GAS FLOW INCREASING   ⟶

Horizontal flow patterns are bubbly, plug, stratified, wavy, slug, and annular flow.

Wavy flow   Slug flow   Annular flow

GAS FLOW INCREASING   ⟶

Fig. 2.2 Flow patterns in horizontal flow in a tube.

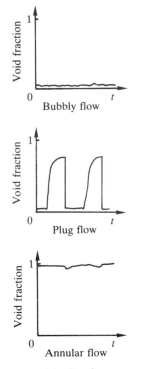

Fig. 2.3 Idealized response of a void-fraction probe to determine the flow pattern.

while others delineate minor differences in the main flow patterns. The number of flow patterns shown in Figs. 2.1 and 2.2 probably represent the minimum which can sensibly be defined. Further general details about flow patterns can be found in Collier and Thome (1994) and Hewitt (1982).

2.2   Flow pattern determination

The main methods of determining the flow pattern are:

1. by eye, but this is subjective and is only possible for flow in transparent tubes; and

2. by various types of instrument, for example, the gamma-ray densitometer which measures mean density across the tube or average void fraction (the proportion of the total volume which is occupied by gas). The ideal response which might be obtained for various flow patterns might be as shown in Fig. 2.3, in which the flow patterns can easily be distinguished. However, the results are rarely so conclusive and so interpretation is again subjective. Such a method does, however, have the advantage that it can give results even for flow in an opaque tube. A much more detailed discussion of flow pattern determination is given by Hewitt (1978).

2.3   Transitions in vertical flow

It is interesting to look at the transition mechanisms at work as one flow pattern gives way to another. Here, only vertical flow is considered.

1. *Bubbly to plug flow.* The normal mechanism for the transition is that, as the gas flow increases, the bubbles get closer together and collisions therefore occur more often (see Fig. 2.4, taken from Radovich and Moissis 1962). Some of the collisions lead to coalescence of bubbles and eventually to the formation of plugs. A large increase in collision frequency at a void fraction of about 0.3 means that the transition tends to occur around this point. However, the transition may occur at a higher void fraction (possibly as high as 0.6) if coalescence is prevented by the presence of a surface active agent. Also, if the liquid flow rate (and therefore the liquid turbulence intensity) is high, large bubbles will be broken up into smaller ones even at high void fractions.

2. *Plug to churn flow.* Because the gas plug rises through the liquid, the gas velocity in the plug is upwards, but the liquid velocity in the thin film around the plug is usually downwards, and so the flow is countercurrent (see Fig. 2.5). As will be seen in Chapter 6, at some definite flow condition the gas velocity will suddenly disrupt the liquid film (the film will 'flood') and Nicklin and Davidson (1962) therefore suggested that the plug flow will break down to give the pulsating, highly unstable churn flow. Annular flow is not

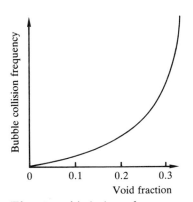

Fig. 2.4 Variation of bubble collision frequency with void fraction.

formed immediately because the gas velocity is not yet high enough to support the film continuously.

3. *Churn to annular flow.* As the gas flow rate (and so the gas velocity) increases, the gas is able to support the liquid as a film on the tube walls and annular flow occurs. This transition is related to the flow reversal point, which in turn is related to flooding. Flow reversal occurs when the gas velocity is reduced and the liquid first starts to flow down instead of up (see also Wallis 1961).

2.4 Flow pattern maps

Flow pattern maps are an attempt, on a two-dimensional graph, to separate the space into areas corresponding to the various flow patterns. Simple flow pattern maps use the same axes for all flow patterns and transitions. Complex maps use different axes for different transitions. Examples of some common, useful flow pattern maps are the follows.

1. The Baker map for horizontal flow (see Fig. 2.6). This map was first suggested by Baker (1954), and was subsequently modified by Scott (1963). The axes are G_g/λ and $G_\ell\psi$, where

$$G_g = \text{mass flux of gas} = \frac{\text{gas mass flow rate}}{\text{tube cross-sectional area}} \qquad (2.1)$$

$$G_\ell = \text{mass flux of liquid} = \frac{\text{liquid mass flow rate}}{\text{tube cross-sectional area}} \qquad (2.2)$$

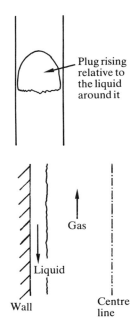

Fig. 2.5 Liquid and gas flows during the passage of a gas plug.

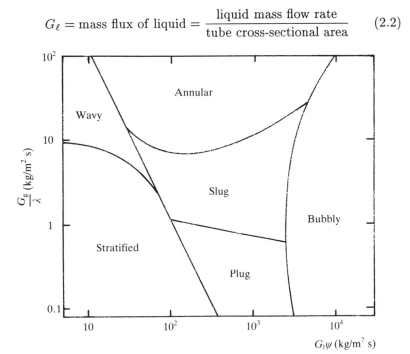

Fig. 2.6 Baker map for horizontal flow in a tube.

ρ_ℓ is the liquid density (kg/m^3); ρ_g is the gas density (kg/m^3); ρ_{water} is the density of water $= 1000$ kg/m^3; ρ_{air} is the density of air $= 1.23$ kg/m^3; μ_ℓ is the liquid viscosity (Ns/m^2); μ_{water} is the viscosity of water $= 10^{-3}$ Ns/m^2; σ is the surface tension (N/m); and σ_{water} is the surface tension of air–water $= 0.072$ N/m.

$$\lambda = \left(\frac{\rho_g}{\rho_{\text{air}}} \frac{\rho_\ell}{\rho_{\text{water}}} \right)^{\frac{1}{2}} \tag{2.3}$$

$$\psi = \frac{\sigma_{\text{water}}}{\sigma} \left(\frac{\mu_\ell}{\mu_{\text{water}}} \left[\frac{\rho_{\text{water}}}{\rho_\ell} \right]^2 \right)^{\frac{1}{3}} \tag{2.4}$$

2. The Hewitt and Roberts (1969) map for vertical upflow in a tube (see Fig. 2.7). Note that here G^2/ρ is a momentum flux, and so all the transitions are assumed to depend on the phase momentum fluxes. Wispy annular flow is a sub-category of annular flow which occurs at high mass flux when the entrained drops are said to appear as wisps or elongated droplets.

The Hewitt and Roberts map works reasonably well for water–air and water–steam systems over a range of pressures, again in small diameter tubes.

For both maps it should be noted also that the transitions between adjacent flow patterns do not occur suddenly but over a range of flow rates. Thus the lines should really be replaced by rather broad transition bands.

3. The Taitel and Dukler (1976) map for horizontal flow. This is the best-known example of the complex type of flow pattern map. The map is shown in Fig. 2.8, and the parameters necessary are

$$X = \left[\frac{(\mathrm{d}p/\mathrm{d}z)_\ell}{(\mathrm{d}p/\mathrm{d}z)_g} \right]^{\frac{1}{2}} \tag{2.5}$$

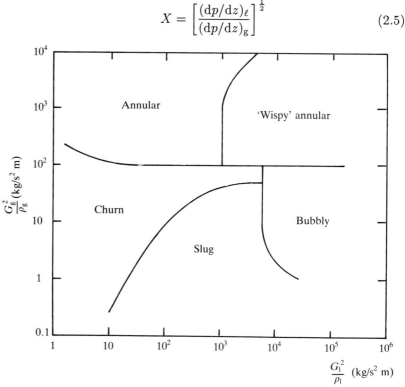

Fig. 2.7 Hewitt and Roberts map for vertical upflow in a tube.

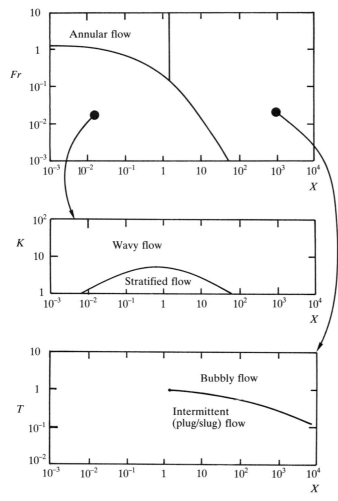

Fig. 2.8 Taitel and Dukler method for flow pattern determination in horizontal flow in a tube.

$$Fr = \text{Froude number} = \frac{G_\text{g}}{[\rho_\text{g}(\rho_\ell - \rho_\text{g})dg]^{\frac{1}{2}}} \qquad (2.6)$$

$$T = \left[\frac{|(\mathrm{d}p/\mathrm{d}z)_\ell|}{g(\rho_\ell - \rho_\text{g})}\right]^{\frac{1}{2}} \qquad (2.7)$$

$$K = Fr \left[\frac{G_\ell d}{\mu_\ell}\right]^{\frac{1}{2}} \qquad (2.8)$$

Note that X is often known as the Martinelli parameter, and that the modulus sign in eqn (2.7) ensures that T is always positive. $(\mathrm{d}p/\mathrm{d}z)_\ell$ = the frictional pressure gradient if the liquid in the two phase flow were flowing alone in the tube (N/m^3),

$(\mathrm{d}p/\mathrm{d}z)_{\mathrm{g}}$ = the frictional pressure gradient if the gas in the two-phase flow were flowing alone in the tube (N/m^3),

d = tube diameter (m),

g = acceleration due to gravity (9.81 m/s^2), and

μ_ℓ = liquid viscosity (Ns/m^2).

The Taitel and Dukler flow pattern map has a better scientific basis than many maps, and often extrapolates well to extreme conditions.

Again, the transition lines should be shown as broad bands. All the transition criteria used by Taitel and Dukler have some theoretical basis, although this is sometimes rather tenuous. For example, the general approach is to take the flow rates of the phases and to work out the liquid depth h_ℓ (see Fig. 2.9) if the flow pattern were perfectly stratified. This is done by evaluating the pressure gradient in each phase and then adjusting h_ℓ to make the two pressure gradients equal. Taitel and Dukler say arbitrarily that intermittent flow will occur if $h_\ell/d > 0.5$, because a sinusoidal wave on the stratified flow will touch the top of the tube before the bottom. Such a crude approach is legitimately open to criticism but the overall method does, in general, produce sensible results capable of application to unusual fluids and conditions. For a critical review of the method, see Hewitt (1982).

A complex type of flow pattern map has also been proposed for vertical flow but it has not been so well tested.

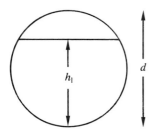

Fig. 2.9 Taitel and Dukler method: idealized stratified flow.

2.5   Other geometries

Flow patterns have been studied in other geometries and generally behave in ways rather similar to the round tube behaviour.

1. Rectangular channels. The flow patterns are very similar to those in round tubes, however liquid tends to collect in the corners.

2. Vertical downflow (in tubes). When both phases are flowing downwards annular flow occurs very frequently (see Barnea *et al.* 1980; Golan and Stenning 1969).

3. Obstructions on the wall. There is a tendency for the liquid to be thrown into the centre of the channel.

4. Helical twisted tape insert. This tends to throw the liquid on to the walls of the tube (so producing good heat transfer).

5. Helically coiled tube (see Fig. 2.10). Again, this tends to throw liquid on to the walls of the tube at point A or point B or somewhere in between (on ACB) due to the effect of gravity. This is discussed further by Whalley (1987).

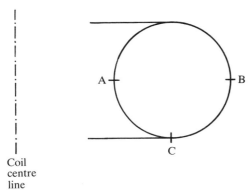

In a helically coiled tube does the liquid tend to gather at A or B?

Fig. 2.10   Helical coil: cross-sectional view.

6. Rod bundles (as in nuclear reactors). The flow patterns are again very similar to those found in round tubes, though of course with many detailed differences. Round-tube flow pattern maps even give reasonable results when used for axial flow along a bundle of tubes (see Bergles 1969).

3 Homogeneous flow

3.1 Introduction

In homogeneous flow the velocities of the phases are assumed to be equal.

A=Cross-sectional area (m²)

Fig. 3.1 Simplified picture of two–phase flow.

Homogeneous flow is a particular model or picture of two-phase flow. In this model it is assumed that the two phases are well mixed and therefore travel with the same actual velocities.

We will first examine the continuity equations (see Fig. 3.1). Now remember that

$$G = \text{overall mass flux (kg/m}^2\text{s)} = \frac{\text{total mass flow rate}}{\text{total cross-sectional area}} \quad (3.1)$$

then a mass balance on the gas phase gives

$$AG_g = AG(1-x) = \rho_g u_g A_g = \rho_g u_g \alpha A \quad (3.2)$$

where:
A = total cross-sectional area (m²);
A_g = average cross-sectional area occupied by the gas phase (m²); and
u_g = actual velocity of the gas phase (m/s).

Similarly for the liquid phase

$$AG_\ell = AG(1-x) = \rho_\ell u_\ell A_\ell = \rho_\ell u_\ell \alpha A \quad (3.3)$$

where:
A_ℓ = average cross-sectional area occupied by the liquid phase (m²); and
u_ℓ = actual velocity of the liquid phase (m/s).

Hence from eqns (3.2) and (3.3)

$$\alpha = \frac{1}{1 + \left(\frac{u_g}{u_\ell}\frac{1-x}{x}\frac{\rho_g}{\rho_\ell}\right)} \quad (3.4)$$

For the homogeneous flow the phase velocities are equal, i.e. $u_\ell = u_g$, and so

$$\alpha = \frac{1}{1 + \left(\frac{1-x}{x}\frac{\rho_g}{\rho_\ell}\right)} \quad (3.5)$$

When ρ_ℓ/ρ_g is large, the void fraction rises very rapidly once the quality rises even slightly above zero, as is illustrated in Fig. 3.2.

One of the main problems in two-phase flow is the calculation of the pressure gradient and homogeneous flow theory does provide answers to this problem (section 3.2). The accuracy of the homogeneous model for pressure gradient calculations is discussed in section 3.4. First, however, single phase pressure gradients are briefly examined.

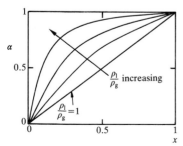

Fig. 3.2 Variation of void fraction with quality in homogeneous flow.

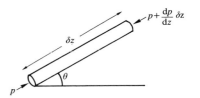

3.2   Single-phase pressure gradient

For flow along the tube illustrated in Fig. 3.3 the momentum equation gives, for steady flow:

pressure force + wall shear force + gravitational force =
change in momentum flow   (3.6)

The terms in eqn (3.6) are

$$-\frac{\mathrm{d}p}{\mathrm{d}z}\delta z \frac{\pi d^2}{4} - \tau \delta z \pi d - \frac{\pi d^2}{4}\delta z \rho g \sin\theta = \frac{\mathrm{d}}{\mathrm{d}z}(Gu\frac{\pi d^2}{4})\delta z \qquad (3.7)$$

Fig. 3.3   Single-phase control-volume for the momentum equation.

where τ = wall shear stress (N/m²). Because $u = G/\rho$, eqn (3.7) can now be arranged into the more convenient form

$$-\frac{\mathrm{d}p}{\mathrm{d}z} = \frac{4\tau}{d} + \rho g \sin\theta + G^2 \frac{\mathrm{d}}{\mathrm{d}z}\left(\frac{1}{\rho}\right) \qquad (3.8)$$

total pressure gradient	=	frictional pressure gradient	+	gravitational pressure gradient	+	accelerational pressure gradient

Thus the total pressure gradient can be expressed as the sum of three components of the pressure gradient. These components arise from distinct physical effects.

The total pressure gradient is made up of three different components.

In single-phase flow, the shear stress is usually expressed in terms of the fraction factor, C_f

$$C_f = \frac{\tau}{\frac{1}{2}\rho u^2} = \frac{\tau}{\frac{1}{2}\frac{G^2}{\rho}} \qquad (3.9)$$

The friction factor is a function of the Reynolds number, Re

$$Re = \frac{\rho u d}{\mu} = \frac{G d}{\mu} \qquad (3.10)$$

and also, in turbulent flow, of the tube roughness ϵ (as in Fig. 3.4).

3.3   Pressure gradient in homogeneous flow

In homogeneous flow, we are treating the two phases as a single fluid with a single velocity. The above analysis for single-phase flow is thus entirely valid. The only remaining questions are:

1. What is the appropriate homogeneous density ρ_h?

2. What is the appropriate homogeneous viscosity μ_h?

The homogeneous density can be found from the phase densities

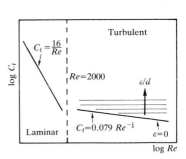

Fig. 3.4   Friction factors in single-phase flow.

$$\rho_h = \rho_g \alpha_h + \rho_\ell(1 - \alpha_h) \tag{3.11}$$

where α_h is the homogeneous void fraction. But this is given by eqn (3.5)

$$\alpha_h = \frac{1}{1 + \left(\frac{1-x}{x}\frac{\rho_g}{\rho_\ell}\right)}$$

So, substituting this into eqn (3.11), we obtain a simple equation for ρ_h

$$\frac{1}{\rho_h} = \frac{x}{\rho_g} + \frac{1-x}{\rho_\ell} \tag{3.12}$$

The equation for homogeneous density ρ_h is unambiguous, but what is the homogeneous viscosity? There is no simple answer.

Equations for the homogeneous viscosity have been largely a matter of guesswork. The suggestions which have produced reasonable results have been given by Isbin *et al.* (1958), Dukler *et al.* (1964), and Beattie and Whalley (1981). The Isbin equation:

$$\frac{1}{\mu_h} = \frac{x}{\mu_g} + \frac{1-x}{\mu_\ell} \tag{3.13}$$

has the obvious attraction of a direct analogy with eqn (3.12) for homogeneous density.

3.4   When does the homogeneous model work well?

This question has to be answered separately for the three components of the total pressure gradient.

1. *Acceleration pressure gradient.* This cannot be measured directly, but the momentum flux can be measured. The homogeneous value seems to give a reasonable prediction, however, the experimental results are rather limited. More details of accelerational pressure gradient and momentum flux are given in Chapter 4.

At suitable conditions, the homogeneous model can give very satisfactory results for the void fraction and the overall pressure gradient. However at low pressures the results can be very inaccurate.

2. *Gravitational pressure gradient.* This, too, cannot be measured directly, but the void fraction can be measured (see Chapter 4). It is found that the homogeneous void fraction is a good estimate of the actual void fraction if $\rho_\ell/\rho_g < 10$ or if $G > 2000$ kg/m^{2}s. If one of these conditions is not met then the homogeneous model can under-predict the mean density by a factor of 5 to 10. It can be noted that for steam-water mixtures the condition that $\rho_\ell/\rho_g < 10$ corresponds approximately to $p > 120$ bar.

3. *Frictional pressure gradient.* This quantity also cannot be measured, but is usually obtained by subtracting the best estimates of the accelerational and gravitational terms from the total experimental pressure gradient. Fortunately it is often found that the frictional term is the dominant one. The homogeneous model, again, gives good results if $\rho_\ell/\rho_g < 10$ or if $G > 2000$ kg/m^{2}s.

3.5   Two-phase multipliers

Two-phase pressure gradients are often expressed in terms of a two-phase multiplier. Thus

two-phase pressure gradient = single phase pressure gradient × two-phase multiplier

Take, for example, the frictional component of the pressure gradient; this can be written as

$$\left(-\frac{dp}{dz}\right)_F = \frac{4\tau}{d} = \frac{4}{d}C_{fh}\frac{1}{2}\frac{G^2}{\rho_h} \tag{3.14}$$

where C_{fh} is the friction factor calculated using ρ_h and μ_h. Now eqn (3.14) can be rewritten as

$$\left(-\frac{dp}{dz}\right)_F = \left[\frac{4}{d}C_{flo}\frac{1}{2}\frac{G^2}{\rho_\ell}\right]\left[\frac{C_{fh}}{C_{flo}}\frac{\rho_\ell}{\rho_h}\right] \tag{3.15}$$

or

$$\left(-\frac{dp}{dz}\right)_F = \left(-\frac{dp}{dz}\right)_{\ell o}\phi_{\ell o}^2 \tag{3.16}$$

where here $(-(dp/dz)_{\ell o}$ is a single phase frictional pressure gradient. The subscript $_{\ell o}$ means:

1. it is a liquid single-phase flow; and

2. it is calculated at a liquid mass flux of G = total mass flux of liquid and gas in two-phase flow.

$\phi_{\ell o}^2$ is the two-phase multiplier, i.e. the factor by which $(-dp/dz)_{\ell o}$ must be multiplied in order to obtain the two-phase frictional pressure gradient.

In a similar way, eqn (3.14) can be rewritten in alternative ways depending on the choice of gas or liquid, and the definition of the flow rate (total mass flow or individual phase mass flow) of the single phase flow. This makes four alternative choices for the two-phase multiplier, see Whalley (1987).

The most convenient are $\phi_{\ell o}^2$ and its analogue ϕ_{go}^2 as these have finite values at qualities of 0 and 1.

Two-phase multipliers are factors by which the single phase (gas or liquid) pressure gradient must be multiplied to get the value for the two-phase pressure gradient.

Or it may be a gas flow.

Or it may be calculated at the flow rate of the individual phase.

Note that two-phase multipliers are usually given the symbol ϕ^2. The fact that ϕ^2 rather than ϕ is used has no special significance: it is only an accident of the past.

4 Pressure drop in two-phase flow— overall methods for separated flow

4.1 Separated flow

In separated flow the phases physically flow with separate, different velocities, as illustrated schematically in Fig. 4.1. From eqn (3.4), the general equation for void fraction α is

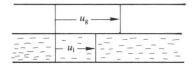

$$\alpha = \frac{1}{1 + \left(\frac{u_g}{u_\ell}\frac{1-x}{x}\frac{\rho_g}{\rho_\ell}\right)} \qquad (4.1)$$

with the velocity ratio u_g/u_ℓ often being called the slip ratio S. Thus, for homogeneous flow S is equal to unity. For separated flow S does not equal unity: it is usually greater than one, so that the gas is moving faster than the liquid phase.

Fig. 4.1 Separated two-phase flow.

4.2 Pressure gradient in separated flow

A comprehensive review of separated flow methods has been provided by Chisholm (1983). Just as in Chapter 3, where the momentum equation was first applied to a single-phase flow, and then to a homogeneous two-phase flow, so it can be applied to the separated flow system. The control volume used is shown in Fig. 4.2.

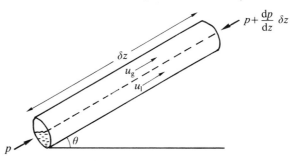

Fig. 4.2 Control volume for momentum equation analysis of separated flow

$$\text{pressure force } + \text{ wall shear force } +$$
$$\text{gravitational force } = \text{ change in momentum} \qquad (4.2)$$

and so

$$-\frac{dp}{dz}\delta z\frac{\pi d^2}{4} - \tau\delta z\pi d - \frac{\pi d^2}{4}\delta z[\alpha\rho_g + (1-\alpha)\rho_\ell]g\sin\theta =$$
$$\frac{\pi d^2}{4}\frac{d}{dz}[\alpha\rho_g u_g^2 + (1-\alpha)\rho_\ell u_\ell^2]\delta z$$

$$(4.3)$$

Now, using the identities

$$u_g = \frac{xG}{\alpha \rho_g} \quad (4.4)$$

$$u_\ell = \frac{(1-x)G}{(1-\alpha)\rho_\ell} \quad (4.5)$$

$$\rho_g u_g^2 = \frac{x^2 G^2}{\alpha^2 \rho_g} \quad (4.6)$$

$$\rho_\ell u_\ell^2 = \frac{(1-x)^2 G^2}{(1-\alpha)^2 \rho_\ell} \quad (4.7)$$

eqn (4.3) becomes

$$-\frac{dp}{dz} = \frac{4\tau}{d} + [\alpha\rho_g + (1-\alpha)\rho_\ell]g\sin\theta + G^2 \frac{d}{dz}\left[\frac{x^2}{\alpha\rho_g} + \frac{(1-x)^2}{(1-\alpha)\rho_\ell}\right] \quad (4.8)$$

It can be noted that:

1. as with homogeneous flow, the total pressure gradient is divided into three components—frictional, gravitation, and accelerational;

2. the gravitational and accelerational terms require a knowledge of the void fraction α;

3. theory has, as yet, provided no help at all with the frictional term. This is often the largest of the three components.

Here the flow was analysed with the momentum equation. An energy equation can alternatively be used. Again the total pressure gradient can be divided into three parts though the division of the total between the frictional, gravitational, and accelerational is different. This point is examined in more detail by Whalley (1987).

Now in separated flow, the phases have different velocities.

Again there are three components of the total pressure gradient: the frictional, gravitational, and accelerational terms.

The energy equation method of analysis leads to complicated and rather confusing equations: the homogeneous density appears in the equations even though the flow is certainly not homogeneous.

4.3 Methods of measuring the void fraction

Experimentally, void fraction can be measured in a number of ways (see, for example, Hewitt 1978).

1. Quick-closing valves. This is a conceptually simple and direct method (see Fig. 4.3). The volumes of liquid and gas trapped when the two valves are simultaneously closed enables the void fraction to be calculated.

$$\alpha = 1 - \frac{\text{volume of liquid trapped}}{\text{total volume between the valves}} \quad (4.9)$$

Quick closing valves are a simple if rather brutal method of measuring void fraction. The method has the great advantage that the results are not open to interpretation.

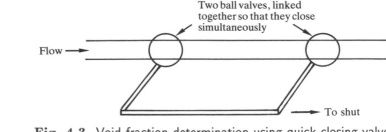

Fig. 4.3 Void fraction determination using quick-closing valves.

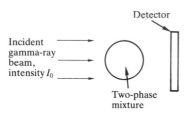

Fig. 4.4 Void fraction determination using a gamma-ray densitometer.

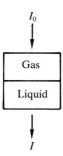

Fig. 4.5 Idealized square channel for gamma-ray densitometer response.

Gamma rays are an attractive non-invasive method, but the radioactive source strengths required are quite high, and even then there is room for doubt because of uncertainties about the void distribution.

2. Gamma-ray densitometer. This instrument is illustrated schematically in Fig. 4.4. For a homogeneous medium (solid metal, for example), then the intensity I of the beam varies like

$$I = I_0 e^{-\mu x} \tag{4.10}$$

where $z =$ distance into the medium (m), and μ is a constant dependent upon the properties of the medium (1/m).

For a flow in a pipe, the intensity of the beam at the detector is measured with the pipe full of liquid (I_ℓ) and then full of gas (I_g). If, in the two-phase flow, the measured intensity is I, then what is the void fraction? We can imagine for simplicity a square pipe. In a stratified flow with the beam as shown in Fig. 4.5

$$\alpha = \frac{\ln(I/I_\ell)}{\ln(I_g/I_\ell)} \tag{4.11}$$

but if the beam were horizontal, then

$$\alpha = \frac{I - I_\ell}{I_g - I_\ell} \tag{4.12}$$

It can be seen, therefore, that the interpretation of the signals depends on the distribution of the phases. This has led to the introduction of densitometers with up to five or more beams which determine the void fraction distribution and so, as well as giving accurate average void fraction, also give information about the flow pattern (see Fig. 4.6). For example, for an annular flow, detectors 1 and 5 in Fig. 4.6 would give low readings because these beams pass obliquely through the liquid film.

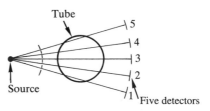

Fig. 4.6 Five-beam gamma-ray densitometer.

4.4 Correlation of void fraction

Most void-fraction correlations are actually correlations of the slip ratio S. Experimentally, it is found that slip ratio depends on (in decreasing order of importance):

1. physical properties (usually expressed as ρ_ℓ/ρ_g);

2. quality x;

3. mass flux G; and then

4. relatively minor variables such as tube diameter, inclination of tube, length, heat flux, and flow pattern.

The dependence of slip ratio on these variables for some common correlations is summarized in Table 4.1. A number of these correlations are discussed briefly below.

Table 4.1
Dependence of slip ratio

Correlation	ρ_ℓ/ρ_g	x	G
Homogeneous model, $S = 1$	no	no	no
Zivi (1964)	yes	no	no
Chisholm (1972)	yes	yes	no
Smith (1971)	yes	yes	yes
CISE (Premoli *et al.*, 1970)	yes	yes	yes
Zuber *et al.* (1967)	yes	yes	yes
Bryce (1977) (steam–water only)	yes	yes	yes

1. Zivi (1964) effectively assumes that total kinetic energy flow is a minimum. The kinetic energy flow in each phase is $\frac{1}{2}\rho_i u_i^2 Q_i$, where Q_i is the volume flow rate of the phase i (m^3/s). Then using the identities

$$Q_g = \frac{GxA}{\rho_g} \tag{4.13}$$

$$Q_\ell = \frac{G(1-x)A}{\rho_\ell} \tag{4.14}$$

and eqns (4.4) and (4.5) for u_g and u_ℓ, the total kinetic energy flow

$$= \frac{1}{2}\rho_g \frac{G^2 x^2}{\alpha^2 \rho_g^2} \frac{GxA}{\rho_g} + \frac{1}{2}\rho_\ell \frac{G^2(1-x)^2}{(1-\alpha)^2 \rho_\ell^2} \frac{G(1-x)A}{\rho_\ell} \tag{4.15}$$

$$= \frac{AG^3}{2}\left[\frac{x^3}{\alpha^2 \rho_g^2} + \frac{(1-x)^3}{(1-\alpha)^2 \rho_\ell^2}\right] = \frac{AG^3}{2}y \tag{4.16}$$

where

$$y = \frac{x^3}{\alpha^2 \rho_g^2} + \frac{(1-x)^3}{(1-\alpha)^2 \rho_\ell^2} \tag{4.17}$$

Now, differentiating y by α to find the minimum kinetic energy flow, we get

$$\frac{dy}{d\alpha} = \frac{-2x^3}{\alpha^3 \rho_g^2} + \frac{2(1-x)^3}{(1-\alpha)^3 \rho_\ell^2} = 0 \tag{4.18}$$

The minimum therefore occurs when

$$\frac{\alpha}{1-\alpha} = \frac{x}{1-x} \left(\frac{\rho_\ell}{\rho_g} \right)^{\frac{2}{3}} \tag{4.19}$$

Comparison with eqn (4.1) gives

$$S = \frac{u_g}{u_\ell} = \left(\frac{\rho_\ell}{\rho_g} \right)^{\frac{1}{3}} \tag{4.20}$$

The slip ratio is therefore assumed to depend only on the phase density ratio.

2. Chisholm (1972) produced a particularly simple correlation, which is

$$S = \left[1 - x \left(1 - \frac{\rho_\ell}{\rho_g} \right) \right]^{\frac{1}{2}} \tag{4.21}$$

Both the Zivi and the Chisholm expressions give $S \to 1$, as $\rho_\ell/\rho_g \to 1$, that is, as the critical point is approached (where the phase densities are equal) the flow becomes homogeneous in character. The Chisholm correlation also gives

$$S \to 1 \quad \text{as} \quad x \to 0$$

and

$$S \to (\rho_\ell/\rho_g)^{\frac{1}{2}} \quad \text{as} \quad x \to 1$$

This latter limit for S is actually the condition that the momentum flow is a minimum.

The performance of the correlations has been assessed in comparison with a wide range of data. The decreasing order of the accuracy in calculating the mean density $\bar{\rho} = (1-\alpha)\rho_\ell + \alpha\rho_g$ is: Bryce (for steam–water only); CISE; Chisholm; Smith; Zuber *et al.*; Zivi; and homogeneous. It can be noted that the Chisholm correlation provides a very simple, reasonably accurate result. The most accurate, generally applicable correlation is the CISE correlation, see Whalley (1987). The standard deviation of the mean density calculated by the CISE correlation is approximately 40%, although it is a little less for steam–water.

4.5   Methods of measuring the momentum flux

The momentum flux has been measured as illustrated in Fig. 4.7. From the momentum equation, because all the momentum in the flow direction is destroyed, the momentum flux (flow per unit area) M (kg/ms^2) is given by

$$M = \frac{F}{A} \qquad (4.22)$$

A large series of experiments has not been performed, but there have been experiments for a reasonable range of variables, for example:

fluid: steam–water
1 bar $< p <$ 10 bar
12 mm $< d <$ 25 mm

One way to plot the results is as the dimensionless number $M\rho_g/G^2$ against the quality x.

1. For single-phase gas flow $(x = 1)$, $M\rho_g/G^2 = 1$.

2. For single-phase liquid flow $(x = 0)$, $M\rho_g/G^2 = \rho_g/\rho_\ell$.

3. The homogeneous model gives

$$M = \frac{G^2}{\rho_h} = G^2 \left[\frac{x}{\rho_g} + \frac{1-x}{\rho_\ell} \right] \qquad (4.23)$$

so

$$\frac{M\rho_g}{G^2} = x + (1-x)\frac{\rho_g}{\rho_\ell} \qquad (4.24)$$

Hence, for homogeneous flow, $M\rho_g/G^2$ is linear with quality. Eqn (4.24) also has the correct limits at $x = 0$ and $x = 1$.

4. The separated model gives an equation which is non-linear with quality, see Whalley (1987). The shape of typical curves for separated flow is illustrated in Fig. 4.8.

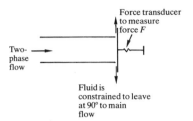

Fig. 4.7 Schematic diagram of apparatus for momentum flux measurement.

Direct measurement of the momentum flux is the only method of directly checking on the methods for calculating accelerational pressure gradient.

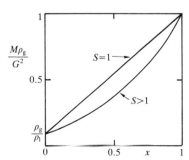

Fig. 4.8 Homogeneous and separated flow; dimensionless momentum flux variation with quality.

4.6   Results for momentum flux

Separated flow gives a lower momentum flux than homogeneous flow.

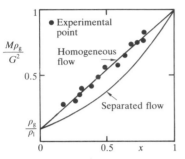

Fig. 4.9 Experimental results for momentum flux compared with the homogeneous and separated flow results.

The homogeneous model gives good results even when the flow is known to be non-homogeneous.

One reason for this is that a two-phase flow is always less steady than a single-phase flow: even a turbulent single phase flow.

Typical results (from Andeen and Griffith 1968) are shown in Fig. 4.9, plotted as suggested in section 4.5. The data points cluster around the homogeneous line. The separated flow line, however, corresponds to a slip ratio known to work well for the calculation of void fraction under the experimental conditions. Why, then, does the separated flow model work badly for momentum flux, but the homogeneous model works well? The flow cannot be both homogeneous and separated at the same time. The answer to this problem is probably that the good prediction of the homogeneous model is a fortuitous accident. It has not been taken into account that the real flow is not steady with time: different phase distributions at different times allow the mass flux to vary (this is most marked in plug flow). Also, there is a velocity profile across the tube (see Fig. 4.10). Larger values of G contribute excessively to the momentum

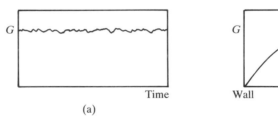

Fig. 4.10 Variation of mass flux with time and position.

flux, which is an area and time average of G^2, hence we are concerned with $\overline{G^2}$ which is larger than $(\overline{G})^2$. Thus, although separated flows may give a good result locally and instantaneously, this averaging problem means that the overall result is too low. Fortuitously, homogeneous flow seems to give a result for the momentum flux of about the correct value.

4.7   Frictional pressure gradient

The frictional pressure gradient can be correlated by various values of ϕ_ℓ^2, $\phi_{\ell o}^2$, ϕ_g^2, and ϕ_{go}^2 as for homogeneous flow, the actual values of the ϕ^2 multipliers usually being determined experimentally. A useful parameter is the Martinelli parameter X^2, which is given by

$$X^2 = \left(\frac{\mathrm{d}p}{\mathrm{d}z}\right)_\ell \bigg/ \left(\frac{\mathrm{d}p}{\mathrm{d}z}\right)_g \qquad (4.25)$$

Note that, as with the various ϕ^2 definitions, X^2 is normally used rather than X. $(\mathrm{d}p/\mathrm{d}z)_\ell$ and $(\mathrm{d}p/\mathrm{d}z)_g$ are the single-phase liquid and gas pressure gradients ($\mathrm{N/m^3}$) respectively, calculated using the actual phase flow rates for the two-phase flow.

It has been found convenient to write

$$\phi_\ell^2 = 1 + \frac{C}{X} + \frac{1}{X^2} \qquad (4.26)$$

where C is a parameter first introduced by Chisholm. Equation (4.26) is equivalent to

$$\phi_g^2 = 1 + CX + X^2 \qquad (4.27)$$

Substitution into either of these equations gives

$$\left(-\frac{dp}{dz}\right)_F = \left(-\frac{dp}{dz}\right)_\ell + C\left[\left(-\frac{dp}{dz}\right)_\ell \left(-\frac{dp}{dz}\right)_g\right]^{\frac{1}{2}} + \left(-\frac{dp}{dz}\right)_g \qquad (4.28)$$

two-phase   liquid term   'two-phase' term   gas term
frictional
pressure
gradient

The magnitude of C can be used to assess how important the specifically two-phase effects are in determining the two-phase frictional pressure drop.

First, we can work out the value of C for homogeneous flow. For the case where the friction factor C_f is a constant, it can be shown that

$$C = \left(\frac{\rho_\ell}{\rho_g}\right)^{\frac{1}{2}} + \left(\frac{\rho_g}{\rho_\ell}\right)^{\frac{1}{2}} \qquad (4.29)$$

A limiting value for C can be calculated for homogeneous flow: where the phases interact so fully they have the same velocity.

This is therefore the value of C for homogeneous flow, and represents an upper limit on C which occurs when the interaction effects between the phases are very large. Note that for atmospheric pressure air-water flow $C = 28.6$.

Another extreme example is a separated flow with no interaction at all between the phases. For fully turbulent flow, when the friction factor C_f is constant, the result is that

$$C = 3.66 \qquad (4.30)$$

Another limiting case is where there is no interaction at all between the phases.

The analysis can be repeated for laminar flow, in which case

$$C = 2 \qquad (4.31)$$

Note that for either turbulent or laminar flow, the values of C are much less than for homogeneous flow.

In actual fact the results show that there is of course always some interaction between the phases even in the most separated flow.

One well-known empirical correlation is that of Lockhart and Martinelli (1949) (see Fig. 4.11). The shaded regions contain four lines depending on whether each phase, flowing alone, would be laminar or turbulent. It has been shown by Chisholm (1967) that the Lockhart–Martinelli curves are well represented by the values of C in Table 4.2. It can be noted that the turbulent–turbulent value for C is much larger than the no-interaction value of 3.66. Hence, in real separated flow, there is considerable interaction between the phases.

Table 4.2
C values of different types of flow

liquid	gas	C value
turbulent	turbulent	20
laminar	turbulent	12
turbulent	laminar	10
laminar	laminar	5

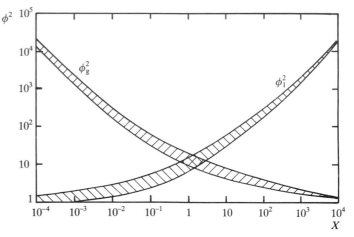

Fig. 4.11 Lockhart–Martinelli frictional pressure drop correlation.

4.8   Practical correlations for frictional pressure gradient

It is not the intention here to write out the detailed form of the correlations, but to give some comments on them.

1. The correlation of Lockhart and Martinelli (1949) (see section 4.7) is simple but not very accurate.

2. Martinelli and Nelson (1948) give values of $\phi_{\ell o}^2$, but only for steam–water flow. Their method is not very accurate.

3. Thom (1964) gives values of ϕ_{lo}^2 but only for steam–water flow above 17 bar. The method gives reasonable results.

4. Baroczy (1966) produced a peculiar graphical correlation for ϕ_{lo}^2 which is applicable to any fluid. Interpolation between the meandering lines on the graphs is difficult but it does give reasonable results.

5. Chisholm (1973) gives an analytical equation of ϕ_{lo}^2 which is applicable to any fluid; however, it is not particularly accurate.

6. Friedel (1979) gives a complicated empirical equation for ϕ_{lo}^2 which is applicable to any fluid, though does not work well when $\mu_\ell/\mu_g > 1000$. It is the best generally available and generally applicable correlation. Even this optimized correlation gives a room-mean-square error of the order of 40%. It seems unlikely that substantial improvements in this accuracy will be made until modelling methods based on individual flow patterns and their characteristics are used.

Even the best empirical methods for the calculation of two-phase pressure gradients give errors of the order of 40%. Recent correlations and large-scale correlation exercises have failed to reduce this figure.

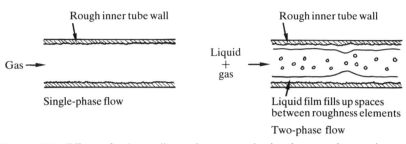

Fig. 4.12 Effect of tube-wall roughness on single-phase and two-phase pressure drop.

4.9 Effect of tube roughness

In single-phase flow when the fluid is turbulent, increasing the tube roughness increases the frictional pressure gradient.

There is some evidence (see Chisholm, 1978) that in two-phase flow the effect of roughness is not as great as in single-phase flow. There is a possible explanation for this (see Fig. 4.12): to the gas in a two-phase flow, the apparent roughness seen is that of the liquid film, not of the wall. It is therefore reasonable that as long as the tube walls are wet, the tube-wall roughness has relatively little effect.

The liquid in the two-phase flow can fill up the crevices and cracks of the rough wall making the tube appear tro be smoother than it really is.

5 Drift flux model

5.1 Introduction

The drift flux model is a type of separated flow model which looks particularly at the relative motion of the phases, and was developed by G. B. Wallis. The most complete reference available to the method is undoubtedly Wallis (1969). It is most applicable to flows where there is a well-defined velocity in the gas phase, for example in bubbly flow and plug flow. It is not particularly relevant to a flow like annular flow which has two characteristic velocities in one phase: the liquid film velocity and the liquid drop velocity. Nevertheless it has been used for annular flows, though with no particular success.

5.2 Development of the model

The model can be confusing due to the large number of connected variables used; it should be noted that u_ℓ and u_g are the actual phase velocities (m/s), V_ℓ and V_g are the superficial phase velocities (m/s) (the superficial velocity is the velocity the phase would have if it were flowing alone in the channel), and u_s is the slip velocity (m/s) $= u_g - u_\ell$. The sign convention for all the velocities is that an upward velocity is positive and a downward velocity is negative.

Definition of slip velocity, u_s.

Now, from the above definition of the slip velocity

$$u_s = \frac{V_g}{\alpha} - \frac{V_\ell}{1-\alpha} \tag{5.1}$$

and so

$$u_s \alpha (1-\alpha) = V_g(1-\alpha) - V_\ell \alpha \tag{5.2}$$

The drift flux $j_{g\ell}$ (m/s) of the gas relative to the liquid is defined by

$$j_{g\ell} = u_s \alpha (1-\alpha) \tag{5.3}$$

Definition of drift velocities, u_{gj} and $u_{\ell j}$.

and the drift velocities, of the gas relative to the mean fluid u_{gj} (m/s), and of the liquid relative to the mean fluid $u_{\ell j}$ (m/s) are defined by

$$u_{gj} = u_g - j \tag{5.4}$$

and

$$u_{\ell j} = u_\ell - j \tag{5.5}$$

where j (m/s) is

$$j = V_\ell + V_g \tag{5.6}$$

It can be seen that the drift velocities are the difference between the actual velocity and the average velocity j. The drift flux is the volumetric flux of a component relative to the surface moving at the average velocity, thus the drift flux of the gas is

$$j_{g\ell} = \alpha u_{gj} = \alpha(u_g - j) \qquad (5.7)$$

Substitution from eqn (5.6) and replacement of αu_g by V_g gives

$$j_{g\ell} = (1 - \alpha)V_g - \alpha V_\ell \qquad (5.8)$$

Similarly, the drift flux of the liquid can be defined as

$$j_{\ell g} = (1 - \alpha)u_{\ell j} = (1 - \alpha)(u_\ell - j) \qquad (5.9)$$

or

$$j_{\ell g} = (1 - \alpha)u_\ell - (1 - \alpha)(V_g + V_\ell) \qquad (5.10)$$

Replacement of $(1 - \alpha)u_\ell$ by V_ℓ then gives

$$j_{\ell g} = -(1 - \alpha)V_g + \alpha V_\ell \qquad (5.11)$$

Thus the two drift fluxes $j_{g\ell}$ and $j_{\ell g}$ are equal and opposite; commonly only $j_{g\ell}$ is used and it is usually called simply the drift flux. In upwards flow with upward velocities being defined as positive, $j_{g\ell}$ is a positive quantity. Homogeneous flow, which is a flow with zero slip velocity, thus corresponds to a flow with $j_{g\ell} = 0$.

5.3 The physical importance of the drift flux

For steady-state one-dimensional flow (see Fig. 5.1) force balances can be written for each phase. For the liquid in the absence of wall shear stress

$$\frac{dp}{dz} + \rho_\ell g - \frac{F}{1 - \alpha} = 0 \qquad (5.12)$$

and for the gas

$$\frac{dp}{dz} + \rho_g g + \frac{F}{\alpha} = 0 \qquad (5.13)$$

Here F is the drag force per unit volume of mixture exerted by one phase on the other. So, eliminating dp/dz from eqns (5.12) and (5.13) we get

$$F = \alpha(1 - \alpha)(\rho_\ell - \rho_g)g \qquad (5.14)$$

In the absence of wall shear, F is a function only of the void fraction, physical properties and the relative motion. Using eqn (5.3), F can be written as

Definition of the drift flux, $j_{g\ell}$.

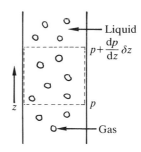

Fig. 5.1 Control volume for force balance.

$$F = \frac{j_{g\ell}}{u_s}(\rho_\ell - \rho_g)g \tag{5.15}$$

Therefore both the drift flux $j_{g\ell}$ and the slip velocity u_s are functions only of α and of the physical properties of the system.

5.4   Example: bubbly flow

For bubbly flow, one equation for u_s, see Whalley (1987), is

This is quite a good empirical equation for the slip velocity of a bubble.

$$u_s = u_b(1 - \alpha) \tag{5.16}$$

where u_b is the rising velocity of a single isolated bubble (m/s). Therefore, using eqn (5.3),

$$j_{g\ell} = u_b\alpha(1 - \alpha)^2 \tag{5.17}$$

This function, eqn (5.17), has been plotted in Fig. 5.2 as a function of the void fraction α, for one particular value of u_b. The graph summarizes the idea that $j_{g\ell}$ is a function only of α and of the system physical properties which determine u_b. It can be noted that since

$$j_{g\ell} = u_s\alpha(1 - \alpha) \tag{5.18}$$

and u_s is always a finite non-zero quantity, then

$$j_{g\ell} \to 0 \quad \text{as} \quad \alpha \to 0$$

and

$$j_{g\ell} \to 0 \quad \text{as} \quad \alpha \to 1$$

We also have a different type of equation for the drift flux

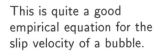
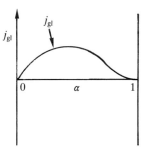

Fig. 5.2 Drift flux (eqn 5.17) plotted as a function of void fraction.

$$j_{g\ell} = (1 - \alpha)V_g - \alpha V_\ell \tag{5.19}$$

which is a consequence of a volume continuity equation for the (assumed) incompressible phases. In eqn (5.19), when:

$\alpha = 0$ then $j_{g\ell} = V_g$,
$\alpha = 1$ then $j_{g\ell} = -V_\ell$,
and $j_{g\ell}$ is linear in α.

Hence another graphical representation of the drift flux $j_{g\ell}$ is shown in Fig. 5.3. The graph summarizes the continuity equation for the system. Combining Figs. 5.2 and 5.3, the two lines for $j_{g\ell}$ intersect; this point gives the actual value of α. The composite graph is shown as Fig. 5.4: essentially it is a graphical solution of eqns (5.17) and (5.19). In Fig. 5.4 the system represented is that of co-current upflows because both superficial velocities (V_g and V_ℓ) are positive. It can be seen from Fig. 5.4 that, for this case, increasing the gas velocity V_g leads to an increase in the void fraction, and increasing the liquid velocity V_ℓ leads to a decrease in the void fraction. Other cases can also be considered.

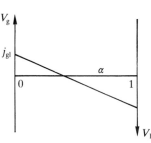

Fig. 5.3 Drift flux (eqn 5.19) plotted as a function of void fraction.

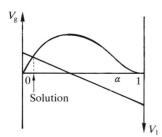

Fig. 5.4 Solution for void fraction for co-current upflow.

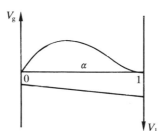

Fig. 5.5 No solution for void fraction for counter-current flow (gas flowing down, liquid flowing up).

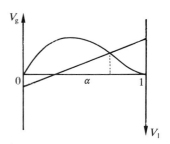

Fig. 5.6 Solution for void fraction for co-current downflow.

1. The liquid flows up and the gas flows down (see Fig. 5.5). Here there is no solution because the situation is not physically possible.

2. The liquid flows down and the gas also flows down (see Fig. 5.6). A solution is obtained.

3. The liquid flows down and the gas flows up (see Fig. 5.7). Here the situation is more complicated. For line 'c', corresponding to a large downward liquid velocity, there is no solution. For line 'a', corresponding to a small downward liquid velocity, there are two solutions; normally the one at lower void fraction is actually obtained. Line 'b' must clearly represent a limit to the counter-current flow. This limit is known as flooding. (A more general description of flooding will be given in Chapter 6.) The possible areas of operation and the flooding locus are shown in Fig. 5.8. The possible values of the superficial velocities at flooding are

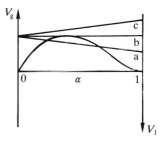

Fig. 5.7 Solution for void fraction for counter-current flow (gas flowing up, liquid flowing down).

$$V_g = u_b 2\alpha^2 (1 - \alpha) \qquad (5.20)$$

and

$$V_\ell = -u_b (1 - 2\alpha)(1 - \alpha)^2 \qquad (5.21)$$

Equations (5.20) and (5.21) satisfy the continuity equation (eqn (5.19)) and produce a line tangential to eqn (5.17). Referring to Fig. 5.8:

at 'A' $V_g = 0$ so $\alpha = 0$; $V_\ell = -u_b$; and
at 'B' $V_\ell = 0$ so $\alpha = \frac{1}{2}$; $V_g = u_b/4$.

Point 'A' corresponds to a dilute suspension of bubbles being held stationary to a downflow of liquid. Point 'B' would not normally be attainable because the bubbles tend to coalesce before a void fraction of 0.5 is reached (see section 2.3).

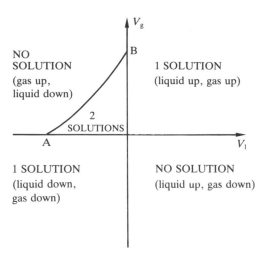

Fig. 5.8   Regions of possible operation and the flooding curve.

5.5   Example: plug flow

The rising velocity of a plug u_p in a tube of diameter d is given (if the tube is not very small), see Whalley (1987), by:

$$u_\mathrm{p} = 0.35(gd)^{\frac{1}{2}} \tag{5.22}$$

and the drift velocity of the gas relative to the mean fluid u_gj is

$$u_\mathrm{gj} = u_\mathrm{p} \tag{5.23}$$

If, however, we now use the result that the plug actually responds to the centre-line velocity, then we should have

$$u_\mathrm{gj} = 0.2j + u_\mathrm{p} \tag{5.24}$$

if the centre-line velocity is 20% greater than the mean velocity. The corresponding drift flux, using eqn (5.7), is

$$j_{\mathrm{g}\ell} = \alpha(0.2j + u_\mathrm{p}) \tag{5.25}$$

Substituting from eqn (5.6) for j, and from eqn (5.8) for $j_{\mathrm{g}\ell}$, then

$$(1 - 1.2\alpha)V_\mathrm{g} - 1.2\alpha V_\ell = \alpha u_\mathrm{p} \tag{5.26}$$

For the particular case of zero liquid flow ($V_\ell = 0$)

$$V_\mathrm{g} = \frac{\alpha u_\mathrm{p}}{1 - 1.2\alpha} \tag{5.27}$$

The equation can also be obtained by other methods, see Whalley (1987). Rearranging eqn (5.26) to obtain a general equation for the void fraction in the presence of a liquid flow as well as a gas flow, we get

$$\alpha = \frac{V_g}{1.2(V_g + V_\ell) + u_p} \tag{5.28}$$

5.6 Corrections due to profile effects

In the previous section an approximate correction method for the effects of the velocity profile was given. Of course, in bubbly flow the void fraction varies across the tube just as the velocity does.

The void fraction could be written as

$$\alpha = \frac{V_g}{u_g} \tag{5.29}$$

Now, since

$$u_g = j + u_{gj} \tag{5.30}$$

eqn (5.29) can be written as

$$\alpha = \frac{V_g}{j + u_{gj}} = \frac{V_g/j}{\left(1 + \frac{u_{gj}}{j}\right)} \tag{5.31}$$

Equation (5.31) can be used as an implicit method of calculating the void fraction. If, for example,

$$j_{g\ell} = u_b \alpha (1 - \alpha)^2 \tag{5.32}$$

then

$$u_{gj} = j_{g\ell}/\alpha = u_b(1 - \alpha)^2 \tag{5.33}$$

and

$$\alpha = \frac{V_g/j}{1 + u_b(1 - \alpha)^2 j} \tag{5.34}$$

However, the effect of velocity profiles and void fraction profiles across the tube has been neglected. In particular,

$$\overline{(\alpha j)} \neq (\overline{\alpha})(\overline{j}) \tag{5.35}$$

Zuber and Findlay (1965) suggested introducing a distribution parameter C_o, which is equal to

$$C_o = \frac{\overline{(\alpha j)}}{(\overline{\alpha})(\overline{j})} \tag{5.36}$$

Zuber and Findlay suggested that for vertical upflow

$$\alpha = \frac{V_g/j}{C_o + u_{gj}/j} \tag{5.37}$$

where $C_o = 1.13$. For bubbly flow they suggested that

$$u_{gj} = 1.4 \left[\frac{\sigma g(\rho_\ell - \rho_g)}{\rho_\ell^2} \right]^{\frac{1}{4}} \tag{5.38}$$

It can be noted that this value of the drift velocity is dependent only upon the physical properties and not upon the void fraction. This would give $j_{g\ell} = \alpha u_{gj}$, and this does not obey the condition that $j_{g\ell} \to 0$ as $\alpha \to 1$. Nevertheless, it gives quite good results in the low-void-fraction region ($\alpha < 0.3$), where bubbly flow actually occurs. Equation (5.38) for u_{gj} is equivalent (for bubbly flow) to putting the rise velocity u_b of a single bubble as

$$u_b = 1.4 \left[\frac{\sigma g(\rho_\ell - \rho_g)}{\rho_\ell^2} \right]^{\frac{1}{4}} \tag{5.39}$$

For low-pressure air–water flow this expression has a value of 0.23 m/s, which is very near the experimental result for equivalent diameters in the range 1 mm to 10 mm, see Whalley (1987). For steam-water flow the value of the bubble rise velocity changes only slowly with pressure, at least in the range 1 bar to 100 bar (see Table 5.1). Near the critical point u_b falls rapidly because $\sigma \to 0$ and $\rho_g / \rho_\ell \to 1$.

Table 5.1

Bubble rise velocity for various pressures of steam–water flow

p (bar)	u_b (m/s)
1	0.22
3	0.21
10	0.20
30	0.19
100	0.16
221.2 (critical)	0

6 Flooding in two-phase flow

6.1 Introduction

The terms 'flooding' and 'flow reversal' are explained with reference to Fig. 6.1. The liquid flow rate is held constant while the gas flow rate is changed as shown. The events are then as follows.

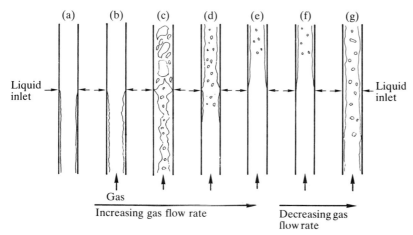

Fig. 6.1 Flooding and flow reversal.

1. First, there is falling film flow with no gas flow. The film has ripples on its surface which grow slowly as the gas flow rate increases.

2. As the rate of upwards flow of gas is increased, the liquid film becomes progressively more disturbed.

3. The liquid film gradually becomes very disturbed with large waves which partially block the channel. Some of the liquid is propelled by the gas flow above the liquid inlet. There are a large number of liquid drops formed in the chaotic flow. This is flooding.

4. More and more of the liquid is pulled upwards by the gas, until

5. all the liquid moves up and co-current annular flow is formed. The liquid film behaves in an oscillatory manner: it moves intermittently up and down, although the net flow is up.

6. The gas flow is now reduced progressively.

7. The liquid film first 'hangs' at the liquid inlet point (see Fig. 6.2). This has been termed the hanging film. The film then begins to flow downwards. This is the flow reversal point.

There is some degree of hysteresis, that is the gas velocity at flooding is greater than that at flow reversal.

Definition of flooding and flow reversal used here.

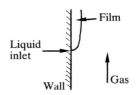

Fig. 6.2 Hanging film.

6.2   Mechanism of flooding

There is no general agreement on the mechanism of flooding, but the simple steady-state analysis given here is certainly too simple.

A number of mechanisms for flooding have been proposed. The most obvious is that the rising gas exerts a shear force on the liquid film; this is analysed using the element of the film shown in Fig. 6.3. If it is assumed that the pressure variation in the gas is negligible, and therefore that the pressure variation in the liquid film in negligible, then a force balance on the film gives

$$\delta z(\tau_i + \tau) = (m - y)\rho_\ell g \delta z \tag{6.1}$$

where the symbols are defined with reference to Fig. 6.3. Note that τ and τ_i are shear stresses (N/m^2). Therefore

$$\tau = (m - y)\rho_\ell g - \tau_i \tag{6.2}$$

Now, if the film is laminar,

$$\tau = \mu_\ell \frac{du}{dy} \tag{6.3}$$

and

$$\frac{du}{dy} = (m - y)\frac{\rho_\ell g}{\mu_\ell} - \frac{\tau_i}{\mu_\ell} \tag{6.4}$$

Integrating,

$$u = \frac{\rho_\ell g}{\mu_\ell}\left[my - \frac{y^2}{2}\right] - \frac{\tau_i y}{\mu_\ell} + C \tag{6.5}$$

where C is the constant of integration. Then $C = 0$ as $u = 0$ when $y = 0$. A special case of eqn (6.5) occurs when the interfacial shear stress τ_i is zero, then

$$u = \frac{\rho_\ell g}{\mu_\ell}\left[my - \frac{y^2}{2}\right] \tag{6.6}$$

and volume flow per unit width in the film Q_ℓ (m^2/s) is given by

$$Q_\ell = \int_0^m u\,dy = \frac{\rho_\ell g}{\mu_\ell}\int_0^m \left(my - \frac{y^2}{2}\right)dy = \frac{\rho_\ell g}{\mu_\ell}\frac{m^3}{6} \tag{6.7}$$

and thus

$$m = \left[\frac{6\mu_\ell Q_\ell}{\rho_\ell g}\right]^{\frac{1}{3}} \tag{6.8}$$

which is the Nusselt film thickness. Returning now to the full equation for u, eqn (6.5), the velocity profile for a fixed film thickness can be plotted as the interfacial shear stress τ_i increases (see Fig. 6.4). Point 'A', where $u = 0$ at $y = m$, occurs when

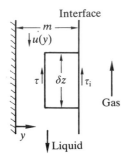

Fig. 6.3 Control volume in the liquid film.

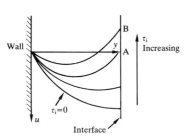

Fig. 6.4 Velocity profiles in a laminar liquid film.

$$\tau_i = \frac{1}{2}\rho_\ell gm \qquad (6.9)$$

Here the interface is stationary. Point 'B', where the net liquid flow is zero, is given by the solution of

$$\int_0^m u \, dy = 0 \qquad (6.10)$$

which is

$$\tau_i = \frac{2}{3}\rho_\ell gm \qquad (6.11)$$

It is tempting to think that this analysis (and point 'B', in particular) will explain flooding. However, the interfacial shear stress τ_i required to initiate flooding is well below that given by any of these formulae (see Hewitt *et al.* 1965)

The most probable mechanism involves the growth of waves on the liquid film. The evidence for this is fourfold: entry phenomena, length effects, wave injection results, and visualization. This evidence is discussed briefly in the following sections.

The true mechanism probably involves waves on the liquid film and their growth.

6.3   Entry phenomena

It is known that the detailed geometry of liquid and gas entry can affect the gas velocity at which flooding occurs. For example, in Fig. 6.5, the flooding gas velocity in apparatus 'A' will be appreciably lower than for apparatus 'B'. The interpretation is that apparatus 'A', because of the sharp-edged gas entry, has a high level of gas-phase turbulence, thus promoting wave growth on the liquid film (see Hewitt 1982).

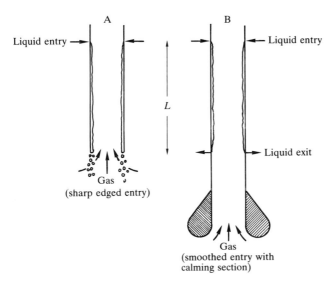

Fig. 6.5 Entry phenomena in flooding.

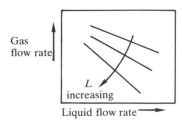

Gas
flow rate

L
increasing

Liquid flow rate

Fig. 6.6 Length effect in flooding.

The experimental evidence on length effects is conflicting, but can be reconciled if the method of introduction of the gas into the tube is considered.

Artifical waves introduced into the liquid film can stimulate flooding to occur, but the position of the injection point for the wave is important.

6.4 Length effects

There has long been disagreement about whether the tube length L has any effect on the gas velocity necessary to initiate flooding. However, it is generally found that in situations with a high turbulence level, such as apparatus 'A' in Fig. 6.5, the tube length has no effect. Where the turbulence level is low, however, such as in apparatus 'B', there is a length effect, as is illustrated in Fig. 6.6. The gas velocity in flooding in long tubes is lower than that in short tubes. The interpretation is that in longer tubes the liquid waves have more time to build up, so the flooding occurs at lower values of the gas velocity.

6.5 Wave injection results

Artificial single waves in co-current upwars annular flow have been shown to behave like naturally occurring waves, see Whalley (1987). Similarly, waves on the falling liquid film can be artificially induced by injecting extra liquid into the film from a syringe. Experiments have been carried out by Whalley and McQuillan (1985), who investigated the effects of injecting waves at positions 'A', 'B', and 'C' in apparatus of the form of Fig. 6.7. The main findings were as follows.

1. Waves injected at 'A' have a large effect, that is they tend to cause flooding easily. Waves injected at 'B', and particularly at 'C', have much less effect.

2. Large injected waves have a greater effect than small waves.

3. Large injected waves cause flooding not at the liquid exit (where normal flooding usually starts) but between the liquid entry and exit.

It seems possible that these findings can all be explained by the wave growth mechanism (see Fig. 6.8). If there is a critical wave amplitude which is necessary to cause flooding and the injected waves grow as they fall, as illustrated in Fig. 6.8, then the three facts above are all illustrated by Fig. 6.8. Flooding is assumed to happen at the points labelled 'x' in this figure, but not at the points labelled 'y'.

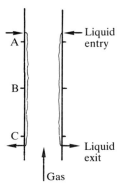

A

B

C

Liquid
entry

Liquid
exit

Gas

Fig. 6.7 Artificial wave injection into a falling liquid film.

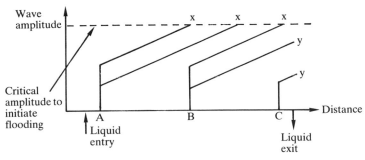

Fig. 6.8 Artificial wave injection and flooding: interpretation of the results.

6.6 Visualization experiments

Two series of photographic experiments have given different views of the flooding process (McQuillan *et al.* 1985).

1. With an axial view down the tube, the explosive growth of a wave at flooding is seen. However, there is little information about the axial position along the tube of any disturbance. An attempt to 'mark' various axial positions in the tube with coloured side light was only partially successful.

2. With a conventional side view, the picture is chaotic so flooding on the inside of an annular test section was photographed (see Fig. 6.9). The central rod was actually a tube with two porous sections of wall to enable film to be formed and removed. The outer transparent shroud tube contained a venturi-like constriction which gives larger air velocities in the region where the camera is focused. This caused flooding to occur reliably within the field of view. The results show waves on the liquid film growing and falling. Just before flooding the waves slow down and then one wave is stationary for a short time. Other waves coalesce with it as they fall into it and the wave thus grows. It is then blown upward by the gas, causing the formation of a large number of liquid droplets.

6.7 Flooding correlations

The methods of correlating flooding which are normally used are of two types.

1. The Wallis (1961) type. The general form of this correlation is

$$V_\ell^{*\frac{1}{2}} + V_g^{*\frac{1}{2}} = C \qquad (6.12)$$

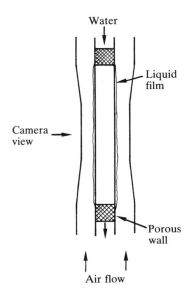

Fig. 6.9 Annulus apparatus for flow visualization.

The waves which appear to cause flooding can be seen in a suitable apparatus.

V_ℓ^* is a dimensionless liquid superficial velocity.

where

V_g^* is a dimensionless gas superficial velocity.

$$V_\ell^* = \frac{V_\ell \rho_\ell^{\frac{1}{2}}}{[gd(\rho_\ell - \rho_g)]^{\frac{1}{2}}} \qquad (6.13)$$

$$V_g^* = \frac{V_g \rho_g^{\frac{1}{2}}}{[gd(\rho_\ell - \rho_g)]^{\frac{1}{2}}} \qquad (6.14)$$

V_ℓ is the liquid superficial velocity (m/s), V_g is the gas superficial velocity (m/s), and C is a constant which has a value of about 0.8.

Equation (6.12) can be represented graphically. The plot is shown in Fig. 6.10 for the particular case when the constant is equal to 0.8. Note that the experimental data tends to lie in a very broad band, and that V_ℓ^* can easily exceed unity.

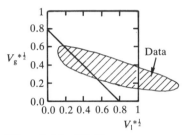

Fig. 6.10 Wallis-type flooding correlation.

2. The Kutateladze number correlation. Alternative dimensionless velocities can be defined using the surface tension, σ, and not the tube diameter d. These are the Kutateladze numbers, for the gas K_g and for the liquid K_ℓ.

$$K_g = \frac{V_g \rho_g^{\frac{1}{2}}}{[g\sigma(\rho_\ell - \rho_g)]^{\frac{1}{4}}} \qquad (6.15)$$

$$K_\ell = \frac{V_\ell \rho_\ell^{\frac{1}{2}}}{[g\sigma(\rho_\ell - \rho_g)]^{\frac{1}{4}}} \qquad (6.16)$$

The most commonly quoted correlation of this type is that of Pushkina and Sorokin (1969)

$$K_g = 3.2 \qquad (6.17)$$

It is sometimes said that to obtain good results use:

1. the Wallis-type correlation if the tube diameter is small (<50 mm), and

However the evidence for this common assertion is far from strong. It does however seem reasonable that in large diameter pipes, the tube diameter should be less likely to be important.

2. the Pushkina and Sorokin correlation if the tube diameter is large (>50 mm).

However, better overall results are obtained with a single, optimized correlation (see McQuillan and Whalley 1985).

$$K_g = 0.286 Bo^{0.26} Fr^{-0.22} \left[1 + \frac{\mu_\ell}{\mu_{\text{water}}}\right]^{-0.18} \qquad (6.18)$$

Bo is the Bond number.

$$Bo = \frac{d^2 g(\rho_\ell - \rho_g)}{\sigma} \qquad (6.19)$$

$$Fr = V_\ell \frac{\pi}{4} d \left[\frac{g(\rho_\ell - \rho_g)^3}{\sigma^3} \right]^{\frac{1}{4}} \qquad (6.20)$$

Fr is the Froude number.

μ_ℓ is the liquid viscosity (Ns/m^2); and μ_{water} is the water viscosity (at room temperature) (Ns/m^2).

Equation (6.18) was developed from a correlation originally devised by Alekseev et al. (1972) for flooding in packed beds.

6.8 Why is the gas Kutateladze number constant?

If we consider flooding to occur when a liquid drop can be supported by the gas flow then, writing a force balance on a drop of diameter D_d, the drag force F is given by

$$F = \frac{\pi D_d^3}{6}(\rho_\ell - \rho_g)g \qquad (6.21)$$

The drag coefficient C_D is defined as

$$C_D = \frac{F/\pi D_d^2}{\frac{1}{2}\rho_g u_g^2} \qquad (6.22)$$

and so

$$F = \frac{1}{8}\rho_g u_g^2 C_D \pi D_d^2 \qquad (6.23)$$

or

$$u_g^2 = \frac{4D_d}{3}\frac{(\rho_\ell - \rho_g)g}{\rho_g C_D} \qquad (6.24)$$

One way of correlating drop size D_d is to use a Weber number We

$$We = \frac{\rho_g u_g^2 D_d}{\sigma} \qquad (6.25)$$

or

$$D_d = \frac{We\,\sigma}{\rho_g u_g^2} \qquad (6.26)$$

Then, eliminating D_d between eqn (6.24) and eqn (6.26) gives

$$u_g^4 = \frac{4}{3}\frac{We}{C_D}\frac{\sigma(\rho_\ell - \rho_g)g}{\rho_g^2} \qquad (6.27)$$

therefore

$$u_g = \left[\frac{4}{3}\frac{We}{C_D}\right]^{\frac{1}{4}}\frac{[\sigma(\rho_\ell - \rho_g)g]^{\frac{1}{4}}}{\rho_g^{\frac{1}{2}}} \qquad (6.28)$$

This simple theory has as its basic assumption that flooding will occur when the gas velocity is large enough to support drops of liquid. Not surprisingly two parameters turn out to be crucial: the diameter of the drops, and their drag coefficient.

or

$$K_{\mathrm{g}} = \left[\frac{4}{3}\frac{We}{C_{\mathrm{D}}}\right]^{\frac{1}{4}} \tag{6.29}$$

Moalem Maron and Dukler (1984) have suggested putting $We = 30$ and $C_{\mathrm{D}} = 0.44$, giving $K_{\mathrm{g}} = 3.1$, which is close to the previously quoted result of Pushkina and Sorokin. However, Hinze (1955) found that for a drop suddenly accelerated $We = 13$, giving $K_{\mathrm{g}} = 2.5$. It is difficult to justify $We = 30$.

It is more difficult to justify the Wallis type of correlation (eqn (6.12)) on theoretical grounds, although Wallis (1969) gives a justification using mixing length theory. However another simple theory does give an interesting result.

6.9   A simple theory based on visualization results

Here, an attempt is made to calculate the gas velocity required to hold the liquid wave stationary. We are thus concerned with a force balance on the wave (see Fig. 6.11). The gas passes through a constriction formed by the wave. If there is no separation before the crest of the wave, then the pressure drop Δp across the wave is given by

$$\Delta p = \frac{1}{2}\rho_{\mathrm{g}}u_{\mathrm{g}}^2\left[1 - \frac{A}{A_1}\right]^2 \tag{6.30}$$

where the areas A and A_1 (see Fig. 6.11), are given by

$$A = \frac{\pi d^2}{4} \tag{6.31}$$

and

$$A_1 = \frac{\pi}{4}(d - 2h)^2 \tag{6.32}$$

Then, if $h/d \ll 1$,

$$\Delta p = \frac{1}{2}\rho_{\mathrm{g}}u_{\mathrm{g}}^2\frac{16h^2}{d^2} \tag{6.33}$$

Now, using the momentum equation and the control volume shown in Fig. 6.11,

$$\Delta p A = (\rho_\ell - \rho_{\mathrm{g}})V_{\mathrm{wave}}g \tag{6.34}$$

where V_{wave} is the volume of the wave which is given by

$$V_{\mathrm{wave}} = \pi h^2\left[\frac{\pi d}{2} - \frac{4h}{3}\right] \tag{6.35}$$

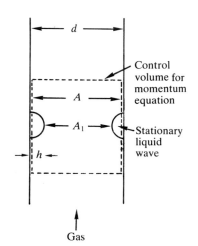

Fig. 6.11 Control volume for momentum equation calculation on a stationary wave.

Note that the effects of the liquid film and the interfacial shear stress have been entirely neglected.

For the case where $h/d \ll 1$,

$$V_{\text{wave}} = \frac{\pi^2 h^2 d}{2} \qquad (6.36)$$

Substituting eqns (6.31), (6.33), and (6.36) into eqn (6.34) gives

$$\frac{1}{2}\rho_g u_g^2 \frac{16h^2}{d^2} \frac{\pi d^2}{4} = (\rho_\ell - \rho_g)\frac{\pi^2 h^2 d}{2} g \qquad (6.37)$$

or

$$\frac{u_g^2 \rho_g}{d(\rho_\ell - \rho_g)g} = \frac{\pi}{4} \qquad (6.38)$$

This is equivalent to

$$V_g^* = \left(\frac{\pi}{4}\right)^{\frac{1}{2}} = 0.89 \qquad (6.39)$$

The final result, eqn (6.39), lies near many experimental results and, in fact, describes the experimental data better than many more complicated theoretical models.

7   Boiling: introduction, pool boiling

7.1   Types of boiling

Boiling can be divided into categories according to the mechanism occurring and according to the geometric situation. The three mechanisms of boiling are:

Mechanisms

1. nucleate boiling, where vapour bubbles are formed (usually at a solid surface);

2. convective boiling, where the heat is conducted through a thin film of liquid—the liquid then evaporates at the vapour–liquid interface with no bubble formation; and

3. film boiling, where the heated surface is blanketed by a film of vapour—the heat is conducted through the vapour, and the liquid vaporizes at the vapour–liquid interface.

The two main geometric situation are:

Geometries

1. pool boiling, where the boiling occurs at a heated solid surface in a pool of liquid which, apart from any convection induced by the boiling, is stagnant; and

2. flow boiling, where the liquid is pumped through a heated channel, typically a tube.

Nucleate boiling and film boiling occur in both pool boiling and flow boiling but convective boiling usually only occurs in flow boiling. First, we look at a classical pool boiling experiment which illustrates nucleate boiling and film boiling.

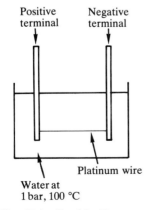

Positive terminal   Negative terminal

Platinum wire

Water at
1 bar, 100 °C

Fig. 7.1 Pool boiling experiment in which the heat flux is controlled.

7.2   Pool boiling

In 1934 Nukiyama performed the experiment illustrated in Fig. 7.1. A platinum wire, immersed in water, was heated electrically. The current in the wire and voltage across the end of the wire enable the power, and therefore the heat flux, to be calculated. Also, from the resistance of the wire the temperature of the wire can be found. The results are illustrated in Fig. 7.2. The curve of the heat flux ϕ against the temperature difference ΔT_{sat} (which is the wall temperature minus the liquid saturation temperature) is the boiling curve. The regions of the curve are:

A to B	natural-convection single-phase liquid—there is no boiling in this region;
B to C	nucleate boiling; and
F to D to E	film boiling.

The important points on the curve are:

B   the onset of nucleate boiling;
C   the burnout point, where the heat flux is equal to the critical heat flux; and
F   the minimum film boiling point.

Note that when, as in the case of the apparatus shown in Fig. 7.1, the heat flux is controlled there is a hysteresis loop. The complete curve (shown by a dotted line in Fig. 7.2) was guessed by Nukiyama. The complete curve can be obtained by controlling the temperature rather than the heat flux. This can be done by, for example, heating the heat transfer surface by means of a hot liquid, as illustrated in Fig. 7.3 (see Bennett *et al.* 1967). The heat flux can be calculated by measuring the temperature drop in the hot liquid. The boiling curve obtained is shown in Fig. 7.4. It can be noted that in this case there is no hysteresis. The part of the boiling curve between C and F is known as the transition boiling region.

7.3   Visualization of events along the boiling curve

Direct visual and photographic evidence shows the following.

1. The nucleate boiling region (BC in Fig. 7.2) consists of two parts:

    (a) the isolated bubble region, where bubbles behave independently (as illustrated in Fig. 7.5(a)); and

    (b) the slugs and columns region, where the bubbles start to merge and to depart from the heated surface by means of jets which then form large bubbles, or slugs, above the surface (see Fig. 7.5(b)).

2. The film boiling region (FDE in Fig. 7.2) is where the heated surface is covered with a layer of vapour (see Fig. 7.5(c)). The liquid is not in contact with the heated surface. The vapour surface is unstable, and bubbles are released from it into the liquid.

3. The transition boiling region (FC in Fig. 7.2) is a complex region where parts of the surface are in film boiling regime and parts in the nucleate boiling regime of the slugs and columns type.

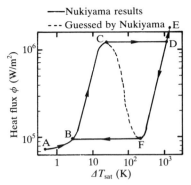

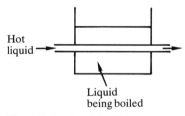

Fig. 7.2 Boiling curve from a heat-flux-controlled surface.

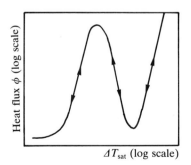

Fig. 7.3 Pool boiling experiment in which the wall temperature is controlled.

Fig. 7.4 Boiling curve from a temperature-controlled surface.

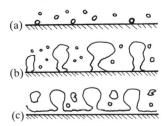

Fig. 7.5 Visualization results in nucleate and film boiling.

7.4   Bubble growth in nucleate boiling

Nucleate boiling can occur when the bulk liquid is saturated (when it is at its boiling temperature) or when it is subcooled (when it is below its boiling temperature). One difference is in what happens to the growing bubbles at the heating surface. In saturated liquid, a bubble grows and then, aided by buoyancy, leaves the surface. As it leaves, fresh liquid flows towards the surface. Another bubble then begins to grow at the same point. In subcooled liquid, the bubble grows and reaches out into the relatively cool liquid: thus the vapour begins to condense, and in doing so causes the liquid temperature to rise slightly. The bubble can collapse completely and, once again, new, cool liquid flows into the area near the wall; the process of bubble growth can then start again.

The collapsing of bubbles in subcooled liquid is responsible for the characteristic 'singing' sound of a kettle when relatively cool water is being heated up to its boiling point. It can also be noted that the bubble is only in contact with a very small area of the heated wall. There is normally no large, dry patch underneath the bubble.

Bubble growth is controlled by two factors.

Bubble growth is limited by inertia forces or by thermal diffusion.

1. The inertia of the liquid. This is relevant at short times after the bubble is formed, and while the inertia is the controlling effect the bubble radius is proportional to the time elapsed since bubble formation.

2. Thermal diffusion through a boundary layer around the bubble. The latent heat for the evaporation must be supplied from the hot liquid layer around the growing bubble. While the thermal diffusion effect is controlling the radius it is proportional to the square root of time elapsed since bubble formation. Thermal diffusion becomes the controlling effect at relatively long times after the bubble starts its growth.

Thus the bubble grows in the matter illustrated in Fig. 7.6. At short times the bubble radius is proportional to t (inertia-controlled). At long times the bubble radius is proportional to $t^{\frac{1}{2}}$ (thermal-diffusion-controlled).

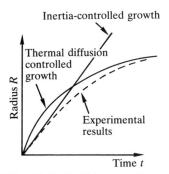

Fig. 7.6 Bubble growth as a function of time.

7.5   Bubble nucleation

The process of bubble formation is known as nucleation. In many ways it is analogous to crystal nucleation. A salt solution can be cooled until it becomes supersaturated, but before a crystal can start growing it must have something to start growing upon; a rough surface, dirt particles or a small crystal. Bubble nucleation of dissolved gas to form gas bubbles in a liquid can be seen in a carbonated drink in a glass. Careful observation will show that the gas bubbles:

1. are formed at a surface on the glass container;

2. rise in a chain of bubbles originating from the same spot on the surface; and

3. often originate from the same spot if the glass is emptied and refilled.

If the liquid is clean (that is it does not contain dirt particles) the bubbles are not observed to arise from a spot inside the bulk of the liquid, away from the surface.

The surface properties are obviously important in all these nucleating phenomena: crystallization, gas bubble formation, and nucleate boiling. The surface, when viewed under a high-power microscope, is not smooth: it contains pits and cracks in a complex pattern. Smoother and smoother surface finishes only make the size scale of the pits and cracks smaller and smaller. A very enlarged cross-section through the surface might be as shown in Fig. 7.7 The cracks and crevices do not, of themselves, constitute nucleation sites for the bubbles: they must also contain pockets of gas, probably air trapped when the vessel was filled with the liquid. It is from these pockets of trapped air that the vapour bubbles begin to grow during nucleate boiling.

Surface properties are important in nucleation phenomena.

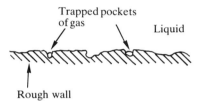

Fig. 7.7 Enlarged view of a boiling surface.

7.6 Types of nucleation

The usual type of nucleation is that described in the previous section: nucleation at a solid surface. This is called heterogeneous nucleation because a solid and a liquid are involved. In extreme cases, where there are no nucleation sites on the vessel walls, homogeneous nucleation may occur. This is nucleation which occurs in the bulk liquid away from the walls of the vessel. Such an extreme situation could be created by the apparatus shown in Fig. 7.8. Here, water is heated while contained in a rotating 'dish' of mercury. If the water and the mercury are clean, there will be no nucleation sites available for bubble growth. The temperature of the water, which is at a pressure of 1 bar, can be raised far above $100\,^\circ\mathrm{C}$ before boiling starts.

Homogeneous nucleation occurs when, by randomly occurring processes, a number of energetic molecules come together to form a small, very localized, 'vapour' region. This vapour region can then grow rapidly. Once homogeneous nucleation begins the vapour bubbles grow at an explosive rate.

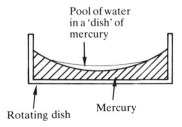

Fig. 7.8 Apparatus for a homogeneous nucleation experiment.

Alternatively homogeneous nucleation can also be studied as a drop of liquid falls or rises through a much hotter immiscible liquid.

For many organic substances, at 1 bar, homogeneous nucleation occurs (see Blander and Katz 1975) at about $0.89T_\mathrm{c}$, where T_c is the critical

For water at 1 bar this rule does not work too well, and the homogeneous nucleation temperature is about 321 °C ($= 0.92T_c$).

temperature (K). The homogeneous nucleation temperature increases as the pressure increases, and becomes equal to the critical temperature when the pressure reaches the critical pressure. The latent heat necessary to fuel the large bubble growth rate is supplied from the very hot liquid.

Heterogeneous nucleation is the most common type of nucleation in practice.

Homogeneous nucleation can be important in some clean systems, using organic liquids, which are operated close to the critical pressure. However, under these conditions the vaporization is much more gentle and less explosive.

7.7   Heterogeneous nucleation in pool boiling

The ideal cavity is pictured (see Fig. 7.9) as being conical and having a circular opening. As the bubble grows its radius of curvature changes.

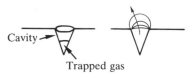

Cavity

Trapped gas

Fig. 7.9 Ideal nucleation cavity.

The minimum radius of curvature of the bubble occurs when the bubble forms a hemisphere at the cavity mouth. The radius of curvature is then, of course, equal to the radius R of the cavity opening.

If the pressure inside the bubble is p_B (N/m^2), then

$$p_B = p + \frac{2\sigma}{r} \qquad (7.1)$$

where p is the pressure in the liquid (N/m^2), and r is the bubble radius (m). Now, p_B is a maximum when $r = R$ (the cavity radius). The wall temperature T_w must be high enough to vaporize the liquid at a pressure of p_B (see Fig. 7.10). so, for the bubble to grow,

$$T_w > T_{sat} + \frac{dT}{dp}(p_B - p) \qquad (7.2)$$

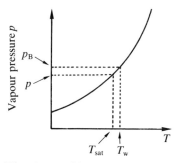

Fig. 7.10 Vapour pressure curve: superheat required for nucleation.

The slope of the vapour pressure curve can be found from the Clausius–Clapeyron equation

$$\frac{dp}{dT} = \frac{\lambda}{(v_g - v_\ell)T_{sat}} \qquad (7.3)$$

Where λ is the latent heat of vaporization (J/kg), T_{sat} is the saturation temperature (K), and v_g and v_ℓ are the specific volumes of the gas and the liquid (m^3/kg). Then if $v_g \gg v_\ell$ and, since $v_g = 1/\rho_g$, then

$$\frac{dT}{dp} = \frac{T_{sat}}{\lambda \rho_g} \qquad (7.4)$$

and the inequality (7.2) becomes

$$T_w > T_{sat} + \frac{2\sigma}{R}\frac{T_{sat}}{\lambda \rho_g} \qquad (7.5)$$

If ΔT_{sat} is the value of $(T_{\text{w}} - T_{\text{sat}})$ at which nucleation starts, then the cavity radius is given by

$$R = \frac{2\sigma T_{\text{sat}}}{\rho_{\text{g}} \lambda \Delta T_{\text{sat}}} \qquad (7.6)$$

For water at 1 bar, ΔT_{sat} is commonly about 5 K so, putting

$$
\begin{aligned}
T_{\text{sat}} &= 373 \text{ K} \\
\sigma &= 0.059 \text{ N/m} \\
\lambda &= 2.256 \times 10^6 \text{ J/kg} \\
\rho_{\text{g}} &= 0.598 \text{ kg/m}^3
\end{aligned}
$$

R is found to be about 6.5 μm, and typical cavity sizes are in the micron range. If the cavity size is known, then clearly the wall superheat ΔT_{sat} required to start nucleate boiling can be calculated. Real surfaces, of course, can contain a range of cavity sizes. As the wall superheat ΔT_{sat} is increased, cavities of a smaller and smaller radius are able to become active and initiate nucleation.

The wall temperature to start nucleation can be found if the size of the cavities is known.

7.8 Heat transfer in nucleate boiling

It is evident that it will be difficult to obtain any general theoretical method of calculating heat transfer coefficients in nucleate boiling. This is because the boiling occurs at nucleate sites, and the number of sites is very dependent upon:

Analytical methods are not available: there are only empirical correlations to rely upon.

1. the physical condition and preparation of the surface; and

2. how well the liquid wets the surface and how efficiently the liquid displaces air from the cavities.

Certainly, equations of the type

$$\phi \propto \Delta T_{\text{sat}}^{1.2} n^{0.33} \qquad (7.7)$$

where n is the nucleation site density (number per unit area), have been produced. However, this is not of much practical use of the variation of n with ΔT_{sat} is not known. Disregarding n, and simply looking at the variation of heat flux with temperature difference, it has been found many times that

Methods which depend on knowing the nucleation site density are interesting academically, but useless in practice.

$$\phi \propto \Delta T_{\text{sat}}^{a} \qquad (7.8)$$

where $a = 3$ to 3.33.

Since we can also write

$$\phi = h \Delta T_{\text{sat}} \qquad (7.9)$$

then

$$h \propto \Delta T_{\text{sat}}^{a-1} \qquad (7.10)$$

and

$$h \propto \phi^{(a-1)/a} \qquad (7.11)$$

where

$$\frac{a-1}{a} = \frac{2}{3} \quad \text{when} \quad a = 3$$

and

$$\frac{a-1}{a} = 0.7 \quad \text{when} \quad a = 3.33$$

The heat transfer coefficient h (W/m²K) is high in nucleate boiling (typically 10 kW/m²K). It is slightly higher for water, and lower for organic liquids. Nucleate boiling is thus a very efficient heat-transfer system.

Do we take into account the surface effects at all?

Two practical approaches to the calculation of nucleate boiling heat transfer are therefore possible.

1. Attempt to take some account of the surface effects.

2. Ignore all surface effects and produce a method which gives a typical heat-transfer coefficient at the particular heat flux for the fluid.

Rohsenow leaves us with the problem of finding the surface–fluid constant C_{sf}.

Rohsenow (1952) included some effects of the surface. He argued that in single-phase convective heat transfer

$$Nu = f(Re, Pr) \qquad (7.12)$$

where the Nusselt number Nu is:

$$Nu = \frac{hL}{\kappa} \qquad (7.13)$$

the Reynolds number Re is:

$$Re = \frac{uL\rho}{\mu} \qquad (7.14)$$

the Prandtl number Pr is:

$$Pr = \frac{\mu C_p}{\kappa} \qquad (7.15)$$

Taking the physical properties as those of the liquid, there are then two problems in trying to use this type of relation for boiling: what is the velocity u and what is the length scale L?

1. The velocity is taken as the liquid velocity towards the surface which is to supply the vapour which is being produced, so

$$u = \frac{\phi}{\lambda \rho_\ell} \tag{7.16}$$

2. The length scale is taken to be

$$L = \left[\frac{\sigma}{g(\rho_\ell - \rho_g)} \right]^{\frac{1}{2}} \tag{7.17}$$

This is related to the most unstable wave on a liquid–vapour interface (for more details see Whalley 1987).

Hence

$$Nu = \frac{h}{\kappa_\ell} \left[\frac{\sigma}{g(\rho_\ell - \rho_g)} \right]^{\frac{1}{2}} \tag{7.18}$$

$$Re = \frac{\phi}{\lambda \rho_\ell} \left[\frac{\sigma}{g(\rho_\ell - \rho_g)} \right]^{\frac{1}{2}} \frac{\rho_\ell}{\mu_\ell} \tag{7.19}$$

and

$$Pr = \frac{\mu_\ell C_{p\ell}}{\kappa_\ell} \tag{7.20}$$

Rohsenow then correlated Nu with Re and Pr, so that

$$Nu = \frac{1}{C_{\mathrm{sf}}} Re^{1-n} Pr^{-m} \tag{7.21}$$

or, rearranging,

$$\frac{C_{p\ell} \Delta T_{\mathrm{sat}}}{\lambda} = C_{\mathrm{sf}} \left[\frac{\phi}{\mu_\ell \lambda} \left(\frac{\sigma}{g(\rho_\ell - \rho_g)} \right)^{\frac{1}{2}} \right]^n \left(\frac{\mu_\ell C_{p\ell}}{\kappa_\ell} \right)^{1+m} \tag{7.22}$$

where, commonly, $n = 0.33$ and $m = 0.7$. Note that $n = 0.33$ is equivalent to $a = 3$ in eqn (7.8). C_{sf} is the surface-fluid constant which depends on both the surface and the fluid. Typical values are between 0.0025 and 0.015. Note that, for a given ΔT_{sat}, the heat flux is proportional to C_{sf}^{-3}. Since C_{sf} can vary by a factor of 10, the heat flux can vary by a factor of 1000.

The heat flux at the onset of nucleate boiling can be found by using eqn (7.5) to find ΔT_{sat} when nucleate boiling starts, and then by using eqn (7.22) to find the corresponding heat flux.

Often the surface effects are unknown and C_{sf} is not known, so Rohsenow's correlation is no help. This has led to the development of methods which ignore all surface effects, and are thus relatively inaccurate.

Other methods like those of Mostinski do not take the surface into account at all.

A typical, and quite effective, overall blanket correlation is that of Mostinski (see Mostinski 1963; and Starczewski 1965).

$$h = 0.106p_c^{0.69}\phi^{0.7}f(p_R) \qquad (7.23)$$

where h is the pool-boiling heat-transfer coefficient (W/m^{2}K), ϕ is the heat flux (W/m^2), p_c is the critical pressure (bar), $f(p_R)$ is a function of the reduced pressure p/p_c, and

$$f(p_R) = 1.8p_R^{0.17} + 4p_R^{1.2} + 10p_R^{10} \qquad (7.24)$$

Equation (7.23) is a dimensional equation and so the units given above must be used. Note that $h \propto \phi^{0.7}$, and that all the other physical properties and their variations with pressure are 'buried' in p_c and p_R. h increases monotonically with p_R, and rises rapidly as the critical pressure ($p_R = 1$) is reached. This type of equation is known to work reasonably well, but why? Cooper (1984) has explained that properties like density, surface tension, and latent heat can be expressed as functions of the reduced variables such as p_R. Moreover, Cooper has derived a new correlation

Cooper compromises by characterising the surface by means of the roughness ϵ, an oversimplification, but at least a parameter which can be measured.

$$h = Ap_R^{(0.12-\log_{10}\epsilon)}(-\log_{10}p_R)^{-0.55}M^{-0.5}\phi^{\frac{2}{3}} \qquad (7.25)$$

where M is the molecular weight (for example, for water $M = 18$) and A is a constant. A best-fit line through the data gave $A = 55$, and a conservative value of A is approximately 30–40. ϵ is the surface roughness in microns. The units of h are (W/m^{2}K) and the units of ϕ are (W/m^2). Equation (7.25) is also a dimensional equation, and so the units given above must be used. Like the Mostinski equation this equation gives values of the heat transfer coefficient which increase with pressure, but some attempt to include surface properties has been made by including the roughness ϵ. As the roughness increases, the heat transfer coefficient also increases.

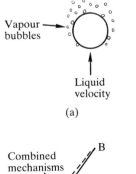

7.9   Effect of liquid velocity on nucleate boiling

So far only pure nucleate boiling has been discussed. What, however, is the effect of superimposing a liquid flow, as in Fig. 7.11(a)? Single-phase forced convection would give a constant heat-transfer coefficient for a given liquid velocity (line 'A') in Fig. 7.11(b), whereas nucleate boiling gives a much steeper line (line 'B'). It seems (see Bergles and Rohsenow 1964; Kutateladze 1969) that the mechanism gradually changes from single-phase forced convection to nucleate boiling as the heater temperature is raised (see dotted line in Fig. 7.11(b)).

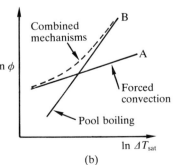

Fig. 7.11   Effect of liquid velocity on nucleate boiling.

At different liquid velocities, the single-phase forced convection heat-transfer coefficient increases with the velocity (see Fig. 7.12), but the nucleate boiling line is unaltered. So at high liquid velocities nucleate boiling only begins to dominate at a comparatively large value of ΔT_{sat}.

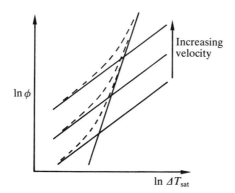

Fig. 7.12 Effect of different liquid velocities on nucleate boiling.

8   Critical heat flux in pool boiling

8.1   Introduction

In pool boiling the critical heat flux occurs when the heated surface is covered with vapour bubbles, and these bubbles form a barrier to incoming liquid. For a heat-flux controlled surface the temperature rise can be very large when the critical heat flux is exceeded (sometimes more than 1000 K).

Classical theory of pool boiling looks at the instability of vapour on a heated surface, and then the instability of the vapour as it leaves the heated surface.

The steps in the classical theory of pool-boiling critical heat flux are (from Lienhard 1981):

1. the instability of the vapour layer on the heated surfaces; and

2. the instability of the vapour rising from the vapour layer through the liquid.

This is particularly interesting because it is an example of a comparatively simple theory which gives excellent results.

8.2   Instability of the vapour layer

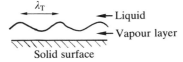

Fig. 8.1 Instability in the vapour layer on a flat plate.

A light fluid in a layer which has a heavy fluid on top of it is unstable. The layer breaks down by the formation of waves on its surface (Fig. 8.1). The wavelength of the most unstable wave (the one with the fastest growth rate) is λ_T. These waves are known as Taylor waves. For a horizontal layer as shown in Fig. 8.1 (see Bellman and Pennington 1954)

$$\lambda_T = C \left[\frac{\sigma}{(\rho_\ell - \rho_g)g} \right]^{\frac{1}{2}} \qquad (8.1)$$

For one-dimensional waves, as on a horizontal rod, $C = 2\pi\sqrt{3}$, and for two-dimensional waves, as on a horizontal flat plate, $C = 2\pi\sqrt{6}$. It can be noted that the length scale used in the development of Rohsenow's correlation for nucleate boiling (see Chapter 7) was $[\sigma/(\rho_\ell - \rho_g)g]^{\frac{1}{2}}$.

A vapour layer underneath a liquid is not stable: the most unstable wave has a wavelength λ_T.

From a horizontal plane surface, the vapour tends to arise as jets or columns arranged on a staggered square lattice, as shown in Fig. 8.2. The spacing of the jets is determined by the peaks in the unstable wave, so the separation of the peaks is λ_T. The two-dimensional result, as shown in Fig. 8.2, is

$$\lambda_T = 2\pi\sqrt{6} \left[\frac{\sigma}{(\rho_\ell - \rho_g)g} \right]^{\frac{1}{2}} \qquad (8.2)$$

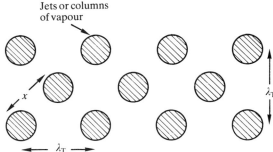

Jets or columns
of vapour

Fig. 8.2 Vapour jets arising from a flat plate.

The vapour rising from a
heated plate or tube rises in
columns at regular intervals.

The dimension x then in Fig. 8.2 is given by

$$x = \frac{\lambda_T}{\sqrt{2}} = 2\pi\sqrt{3}\left[\frac{\sigma}{(\rho_\ell - \rho_g)g}\right]^{\frac{1}{2}} \qquad (8.3)$$

which is also the one-dimensional result. These results then describe
the formation of the vapour jets at the peak amplitude position on the
Taylor waves.

8.3   Instability of the vapour jets

A parallel-sided jet is potentially unstable in the configuration shown in
Fig. 8.3(a): to see this consider the random thinning of the jet which is
illustrated in Fig. 8.3(b).

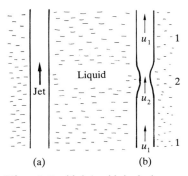

Fig. 8.3 Kelvin–Helmholtz
instability in the vapour jet.

By continuity, $u_2 > u_1$ and therefore, from Bernoulli's equation, $p_2 < p_1$.
If the jet is in equilibrium at 1, then the liquid pressure at 2 will push
the 'neck' further in and disrupt the jet completely, thus breaking it
up. This is a Kelvin–Helmholtz instability. This argument would imply
that the vapour jet is always unstable, but the effects of surface tension,
which has a stabilizing effect, have been neglected.

The jets rising from the
heated surface are than
subject to Kelvin–Helmholtz
instabilities.

If the velocity of the vapour in the jet is u_g, then the dynamic pressure
available is $\frac{1}{2}\rho_g u_g^2$. If the wavelength of the disturbance is λ_{KH} (as shown
in Fig. 8.4), then the radius of curvature of the surface (in the vertical
direction) is proportional to λ_{KH}, and the pressure difference across this
curved surface is proportional σ/λ_{KH}. So when

$$\frac{\sigma}{\lambda_{KH}} \leq A\frac{1}{2}\rho_g u_g^2 \qquad (8.4)$$

where A is an unknown constant, the surface tension will no longer be
able to stabilize the jet. Rearranging, the condition becomes

$$u_g \geq \left(\frac{2\sigma}{A\rho_g\lambda_{KH}}\right)^{\frac{1}{2}} \qquad (8.5)$$

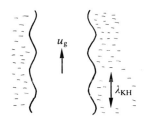

Fig. 8.4 Kelvin–Helmholtz
wavelength.

In fact, a full analysis shows (see Turner 1973) that

$$u_g \geq \left(\frac{2\pi\sigma}{\rho_g \lambda_{KH}} \right)^{\frac{1}{2}} \qquad (8.6)$$

8.4   Calculation of the critical heat flux

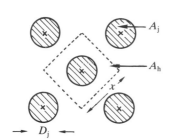

Consider an area of the horizontal plate which supplies one jet (Fig. 8.5) This area is A_h and the jet cross-sectional area is A_j. Then, at the critical heat flux ϕ_c, the rate of heat supply to the area A_h is

$$\phi_c A_h = \lambda \rho_g u_g A_j \qquad (8.7)$$

and so

$$\phi_c = \lambda \rho_g u_g \frac{A_j}{A_h} \qquad (8.8)$$

Fig. 8.5 Vapour yet diameter.

Now, see Fig. 8.5,

$$A_h = x^2 \qquad (8.9)$$

and D_j was first guessed, and then confirmed to be $x/2$, therefore $A_j = \pi(x/4)^2$ and

$$\frac{A_j}{A_h} = \frac{\pi \left(\frac{x}{4}\right)^2}{x^2} = \frac{\pi}{16} \qquad (8.10)$$

We now have enough information to calculate the critical heat flux knowing the spacing of the vapour columns, and when these vapour columns become unstable.

Substituting eqn (8.10) into eqn (8.8),

$$\phi_c = \lambda \rho_g u_g \frac{\pi}{16} \qquad (8.11)$$

where, at the critical condition from eqn (8.5)

$$u_g = \left(\frac{2\pi\sigma}{\rho_g \lambda_{KH}} \right)^{\frac{1}{2}} \qquad (8.12)$$

Then, making the assumption that

$$\lambda_{KH} = x \qquad (8.13)$$

and that

$$x = 2\pi\sqrt{3} \left[\frac{\sigma}{(\rho_\ell - \rho_g)g} \right]^{\frac{1}{2}} \qquad (8.14)$$

and then, substituting eqns (8.12) to (8.14) into eqn (8.11), we get

$$\phi_c = \lambda \rho_g \frac{\pi}{16} \left(\frac{2\pi\sigma}{\rho_g} \right)^{\frac{1}{2}} \left[\frac{(\rho_\ell - \rho_g)g}{\sigma} \right]^{\frac{1}{4}} \left(\frac{1}{2\pi\sqrt{3}} \right)^{\frac{1}{2}} \qquad (8.15)$$

or, simplifying,

$$\phi_c = \frac{\pi}{16 \times 3^{\frac{1}{4}}} \lambda \rho_g^{\frac{1}{2}} [\sigma(\rho_\ell - \rho_g)g]^{\frac{1}{4}} \qquad (8.16)$$

The numerical constant $\pi/(16 \times 3^{\frac{1}{4}})$ has the value of 0.149.

This formula for critical heat flux has been derived in a number of different ways; each time the same formula has been obtained, although with a different constant. Zuber's (1959) original analysis gave a value for the constant of $\pi/24 = 0.131$. The assumption inherent in eqn (8.13), that the Kelvin–Helmholtz wavelength is the same as the Taylor one-dimensional wavelength, can, in fact, be proved to be true.

It will be evident that the equation for the critical heat flux is reminiscent of the Kutateladze number criterion that flooding occurs when

$$K_g = V_g \rho_g^{\frac{1}{2}} [g\sigma(\rho_\ell - \rho_g)]^{-\frac{1}{4}} \qquad (8.17)$$

where V_g is the superficial vapour velocity above the plate (m/s). If now we put

$$\phi_c = \lambda \rho_g V_g \qquad (8.18)$$

then

$$\phi_c = K_g \lambda \rho_g^{\frac{1}{2}} [\sigma(\rho_\ell - \rho_g)g]^{\frac{1}{4}} \qquad (8.19)$$

Kutateladze (1948) performed experiments to determine K_g.

Under reasonable conditions, the final equation for critical heat flux is surpisingly accurate.

8.5 Experimental results

For flat, horizontal plates the equation

$$\phi_c = 0.149 \lambda \rho_g^{\frac{1}{2}} [\sigma(\rho_\ell - \rho_g)g]^{\frac{1}{4}} \qquad (8.20)$$

works very well as long as two conditions are satisfied (see Lienhard and Dhir 1973).

1. Liquid is prevented from entering around the sides of the plate, as it does in Fig. 8.6(a). This can be ensured by adding sides to the flat plate (see Fig. 8.6(b)).

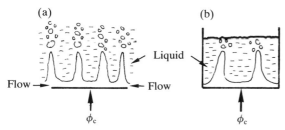

Fig. 8.6 Critical heat flux on a flat plate: conditions for theory to work well.

2. The test section is reasonably large. If the test section dimensions become small then the number of jets to be fitted in becomes important. This makes ϕ_c vary substantially from the predictions of eqn (8.20) when the length is less that $3x$. This is illustrated in Fig. 8.7.

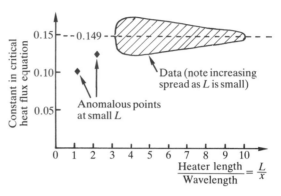

Fig. 8.7 Critical heat flux on a flat, horizontal plate: experiment and theory.

The theory can also be adapted to a cylindrical geometry: industrially this is more important than flat plates.

8.6 Critical heat flux for boiling outside horizontal cylinders

Horizontal cylinders are an important geometry. The basic theory works well, although the numerical constant in eqn (8.20) needs some adjustment because of this geometry. It is found that for large cylinder radii

$$\phi_c = 0.116\lambda\rho_g^{\frac{1}{2}}[\sigma(\rho_\ell - \rho_g)g]^{\frac{1}{4}} \qquad (8.21)$$

For small radii the constant is larger, and is illustrated in Fig. 8.8. The constant can be expressed as (see Sun and Lienhard 1970)

$$\text{numerical constant} = 0.116 + 0.3e^{-3.44R'^{\frac{1}{2}}} \qquad (8.22)$$

where the dimensionless radius R' is defined as

$$R' = R\left[\frac{\sigma}{g(\rho_\ell - \rho_g)}\right]^{-\frac{1}{2}} \qquad (8.23)$$

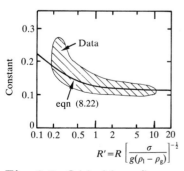

Fig. 8.8 Critical heat flux for boiling on a cylinder: variation with cylinder radius.

and R is the actual radius of the cylinder (m). If $R' > 1$, then the cylinder is effectively large. For water at 1 bar, a dimensionless radius of unity corresponds to an actual radius of 2.5 mm or a diameter of 5 mm. Thus, for tubes of industrial size, the numerical constant in eqn (8.21) can be taken to be 0.116.

8.7 Variation of critical heat flux with pressure

As the system pressure rises:

λ falls slowly at first and then falls steeply as the critical point is approached,

ρ_g increases monotonically,

σ falls monotonically, and

$\rho_\ell - \rho_g$ falls monotonically.

For steam–water, values of the critical heat flux for a flat, horizontal plate calculated from eqn (8.20) are given in Table 8.1.

Note that the maximum critical heat flux occurs at about 70 bar. The variation with pressure is illustrated in Fig. 8.9. This can be compared with the flow-boiling critical heat flux variation with pressure for water given in Chapter 9. For both pool boiling and flow boiling the maximum critical heat flux occurs at about 70 bar.

Table 8.1
Values of critical heat for a flat, horizontal plate using steam–water

p (bar)	ϕ_c $(\mathrm{MW/m^2})$
0.01	0.168
0.1	0.471
1	1.25
10	2.97
30	4.03
50	4.38
70	4.45
90	4.34
100	4.10
150	3.27
221 (p_c)	0

8.8 Effect of liquid subcooling

All the previous results have been for a saturated liquid. If the liquid is cooled an amount ΔT_{sub} (K) below its saturation temperature, then the critical heat flux increases, and the increase is roughly proportional to ΔT_{sub}. One suggestion concerning this has been that made by Ivey and Morris (1962):

$$\phi_{c,\mathrm{sub}} = \phi_{c,\mathrm{sat}} \left[1 + 0.1 \left(\frac{\rho_\ell}{\rho_g} \right)^{\frac{3}{4}} \frac{C_{p\ell} \Delta T_{\mathrm{sub}}}{\lambda} \right] \tag{8.24}$$

$\phi_{c,\mathrm{sub}}$ is the critical heat flux (W/m^2) when the liquid is cooled ΔT_{sub} below its saturation temperature and $\phi_{c,\mathrm{sat}}$ is the critical heat flux (W/m^2) in saturated liquid. Substitution of typical values into eqn (8.24) show that even a moderate degree of subcooling increases the critical heat flux substantially.

For steam–water, the critical heat flux reaches a maximum at about 70 bar. At higher pressures it falls rapidly and becomes zero at the critical pressure.

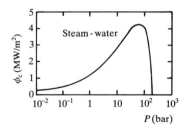

Fig. 8.9 Variation of critical heat flux on a flat, horizontal plate with pressure for steam–water.

8.9 Effect of liquid velocity

For small cylinders, of diameter less than 1 mm (effectively thin wires), the presence of an upward liquid velocity replaces the jet structure by a two-dimensional sheet, as shown in Fig. 8.10 (see Lienhard and Eichhorn 1976).

When the liquid velocity is sufficient to form the two-dimensional wake structure, then the critical heat flux is increased by the presence of the liquid velocity. At high liquid velocities the critical heat flux is proportional to the liquid velocity.

An imposed liquid velocity can completely alter the structure of the flow, and destroy the regular pattern of ascending vapour columns.

For larger cylinders, of diameter greater than 15 mm (that is tubes, see McKee and Bell 1969), there is less evidence, but the critical heat flux does not seem to be greatly affected by the liquid velocity.

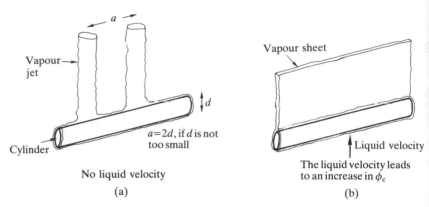

Fig. 8.10 Change in flow structure as a result of an imposed upward liquid velocity. (a) Flow structure with no imposed upward liquid velocity. (b) Flow structure with imposed upward liquid velocity.

9   Flow boiling: onset of nucleation and heat transfer

9.1   Introduction

Flow boiling typically takes place in a vertical boiler tube. The appearance of the boiling mixture for this case is shown in Fig. 9.1. Referring to Fig. 9.1, the distinct areas of the flow are (see Collier and Thome 1994):

1. single-phase liquid,

2. bubbly flow,

3. plug flow

4. churn flow,

5. annular flow,

6. dispersed-drop flow, and

7. single-phase vapour.

The important points in the flow are, again referring to Fig. 9.1, as follows.

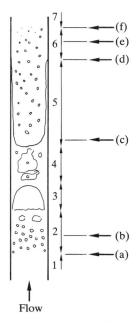

Flow

Fig. 9.1 Flow boiling in a vertical tube.

1. Bubble nucleation begins when the thermodynamic quality x_T is less than zero. The thermodynamic quality is given by

$$x_T = \frac{h - h_{\ell,\text{sat}}}{\lambda} \qquad (9.1)$$

where h is the enthalpy of the flowing mixture (J/kg), and $h_{\ell,\text{sat}}$ is the saturation enthalpy of the liquid phase (J/kg). Thus the nucleation process starts when the liquid in the bulk flow (away from the walls) is still subcooled.

2. $x_T = 0$.

3. Bubble nucleation ceases and the boiling process become convective in character.

4. The liquid film dries out. This is the dryout, burnout, or critical heat flux condition which is discussed separately in Chapter 10.

5. $x_T = 1$.

6. The last liquid drops evaporate.

The facts that bubble nucleation starts when $x_T < 0$ and liquid drops persist when $x_T > 1$ demonstrate that there is no thermodynamic equilibrium in these areas: both the liquid and vapour are not saturated and so not in equilibrium with each other. The temperature variation of the tube wall and the fluid up the tube can be plotted as in Fig. 9.2.

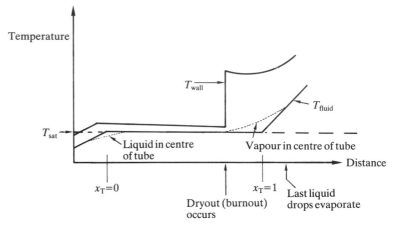

Fig. 9.2 Wall and fluid temperature in flow boiling.

From Fig. 9.2 it can be seen that there is a large increase in wall temperature when dryout occurs: this increase may be many hundreds of degrees. Also, before dryout occurs the difference between wall temperature and the saturation temperature is gradually decreasing. Therefore, the boiling heat transfer coefficient is increasing; this is a consequence of the liquid film in annular flow becoming thinner as the quality increases. One convenient way of expressing the heat transfer coefficient variation along the tube is to plot it against the thermodynamic quality, as in Fig. 9.3. In this figure, lines '1', '2', and '3' represent different heat fluxes. At the highest heat flux, curve '1', burnout occurs before x_T is equal to zero. At a lower heat flux, curve '2', burnout occurs after $x_T = 0$. In the region of positive quality nucleate boiling occurs, and this ceases at burnout. At the lowest heat flux, curve '3', there are five distinct regions. These are (starting from $x_T < 0$):

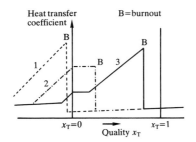

Fig. 9.3 Variation of heat transfer coefficient with quality in flow boiling.

Regions of boiling.

1. single-phase liquid forced convection (in which the heat transfer coefficient is almost constant);

2. subcooled boiling, in which the heat transfer coefficient increases as the bulk fluid approaches the saturation temperature;

3. saturated nucleate boiling in which the heat transfer coefficient is almost constant;

4. saturated convective boiling in which the heat transfer coefficient increases slowly; and

5. post-burnout heat transfer, in which the heat transfer coefficient is low—this regime gradually merges into a single-phase vapour forced-convection regime.

Regimes and types of flow during boiling.

9.2   Onset of nucleate boiling

As in pool boiling, see Chapter 7, the wall temperature must be some way above the saturation temperature before bubble growth can occur. In flow boiling the situation is complicated by the fact that there is a temperature profile in the liquid near the wall (see Fig. 9.4). There is also, in general, a large range of cavity sizes available. As the bubbles grow from these cavities they protrude into the cooler liquid away from the wall. If we restrict our attention to the thin layer next to the wall where heat conduction through the liquid is dominated by molecular thermal conductivity rather than by turbulent convection, then the temperature T (K) near the wall is given by

As in pool boiling, we are concerned with the onset of nucleation. However now because of the flow there is a temperature gradient near the heated surface: going into the flow from the wall the liquid temperature falls. Very near the wall, it can be assumed that the temperature variation is controlled by molecular thermal conductivity.

$$T = T_{\mathrm{w}} - \frac{\phi y}{\kappa_{\ell}} \tag{9.2}$$

where T_{w} is the wall temperature (K), ϕ is the heat flux (W/m^2) from the wall into the fluid, y is the distance from the wall (m), and κ_{ℓ} is the liquid thermal conductivity (W/mK). From Chapter 7, the condition for the bubble of radius y to grow is that

$$T > T_{\mathrm{sat}} + \frac{2\sigma}{y}\frac{T_{\mathrm{sat}}}{\lambda \rho_{\mathrm{g}}} \tag{9.3}$$

Equations (9.2) and (9.3) can be plotted together as in Fig. 9.5. Here, the straight lines are eqn (9.2) plotted at a number of heat fluxes, and the curved line is eqn (9.3).

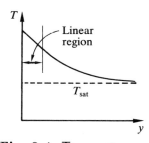

Fig. 9.4 Temperature variation with distance from the heated wall during flow.

One suggestion is that boiling can begin when the curves touch tangentially as at 'A' in Fig. 9.5. At this position the liquid at the outer point of the hemispherical bubble is just hot enough for the bubble to continue growing. At 'A', eqns (9.2) and (9.3) give the same temperature, so

$$T_{\mathrm{w}} - \frac{\phi y}{\kappa_{\ell}} = T_{\mathrm{sat}} + \frac{2\sigma}{y}\frac{T_{\mathrm{sat}}}{\lambda \rho_{\mathrm{g}}} \tag{9.4}$$

or, rearranging,

$$T_{\mathrm{w}} - T_{\mathrm{sat}} = \frac{1}{y}\left(\frac{2\sigma T_{\mathrm{sat}}}{\lambda \rho_{\mathrm{g}}} + \frac{\phi}{\kappa_{\ell}}y^2 \right) \tag{9.5}$$

Also, the temperature gradients are equal, so

$$-\frac{\phi}{\kappa_{\ell}} = -\frac{2\sigma}{y^2}\frac{T_{\mathrm{sat}}}{\lambda \rho_{\mathrm{g}}} \tag{9.6}$$

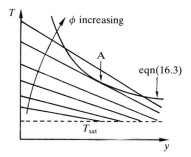

Fig. 9.5 Equations for temperature plotted against distance from the wall.

or

$$y^2 = \frac{2\sigma T_{\text{sat}} \kappa_\ell}{\phi \lambda \rho_g} \tag{9.7}$$

The intersection of the straight lines (the fluid temperature near the heated surface) and the curved line (the nucleation equation) gives the conditions for the onset of nucleate boiling.

Substituting for y from eqn (9.7) into eqn (9.5) gives

$$T_{\text{w}} - T_{\text{sat}} = \left(\frac{\phi \lambda \rho_g}{2\sigma T_{\text{sat}} \kappa_\ell}\right)^{\frac{1}{2}} \left(\frac{2\sigma T_{\text{sat}}}{\lambda \rho_g} + \frac{2\sigma T_{\text{sat}}}{\lambda \rho_g}\right) \tag{9.8}$$

or

$$(T_{\text{w}} - T_{\text{sat}})^2 = \frac{8\sigma T_{\text{sat}} \phi}{\lambda \rho_g \kappa_\ell} \tag{9.9}$$

Thus nucleate boiling will occur if

$$
\begin{aligned}
\sigma &= 0.059 \text{ N/m,}\\
T_{\text{sat}} &= 373 \text{ K,}\\
\lambda &= 2.256 \times 10^6 \text{ J/kg,}\\
\rho_g &= 0.598 \text{ kg/m}^3 \text{, and}\\
\kappa_\ell &= 0.681 \text{ W/mK.}
\end{aligned}
$$

$$\Delta T_{\text{sat}} = \left[\frac{8\sigma T_{\text{sat}} \phi}{\lambda \rho_g \kappa_\ell}\right]^{\frac{1}{2}} \tag{9.10}$$

This equation was first derived by Davis and Anderson (1966).

As an example, calculate ΔT_{sat} for water at 1 bar, for $\phi = 10^5$ W/m^2. From eqn (9.10) using the appropriate physical properties:

$$\Delta T_{\text{sat}} = \left[\frac{8\sigma T_{\text{sat}} \phi}{\lambda \rho_g \kappa_\ell}\right]^{\frac{1}{2}} = 4.38 \text{ K}$$

and the cavity radius operating is given by eqn (9.7)

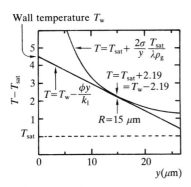

$$R = \left[\frac{2\sigma T_{\text{sat}} \kappa_\ell}{\phi \lambda \rho_g}\right]^{\frac{1}{2}} = 15 \text{ } \mu\text{m}$$

Fig. 9.6 Temperature profiles for the example.

In pool boiling in Chapter 7, it was found that for water at 1 bar, if it was assumed that $\Delta T_{\text{sat}} = 5$ K, then $R = 6.5$ μm. In the current situation the cavity size is much greater because of the effect of the temperature gradient in the liquid (see Fig. 9.6). At the outer edge of the bubble the temperature is half-way between the wall temperature and the saturation temperature.

This nucleation theory, first fully developed by Davis and Anderson (1966), will only work if the following are true.

Conditions for the Davis and Anderson analysis to work well are that all cavity sizes are available, and that the liquid temperature profile is linear.

1. There is a full range of cavity sizes. It does not work, for example, in liquid metals which are very well-wetting and thus very efficient at expelling air from the cavities.

2. The temperature distribution in the liquid is indeed linear. Eventually, of course, the temperature distribution joins evenly with the bulk temperature as shown in Fig. 9.4.

9.3 Nucleate and convective boiling

It is commonly, but not universally, believed that there are, as assumed up to now, two types of behaviour in flow boiling.

1. Nucleate boiling—in which bubbles are formed by nucleation at the solid surface. In highly subcooled boiling these bubbles rapidly collapse, transferring their latent heat to the liquid phase and thus heating it up towards the saturation temperature.

2. Convective boiling—in which heat is transferred by conduction and convection through a thin liquid film. Evaporation then takes place at the liquid–vapour interface.

It is often assumed that only one of the boiling types occurs at once, and that at some point the mechanism suddenly switches from one type of boiling to the other. In fact, the mechanisms can coexist and, as the quality increases, convective boiling gradually supplants nucleate boiling.

9.4 Calculation of flow-boiling heat transfer coefficients

It is now clear that a good procedure for calculating flow-boiling coefficients must have some elements of a nucleate-boiling calculation and some elements of a flow-boiling calculation. This was provided by Chen (1963), whose method (or correlation) is probably the best available and is certainly the best tested. He suggested that the two mechanisms work in parallel, and so

The Chen correlation for boiling heat transfer is the best tested calculation method and provides good answers in many siuations.

$$h_{\mathrm{B}} = h_{\mathrm{NB}} + h_{\mathrm{FC}} \tag{9.11}$$

where h_{B} is the total boiling heat transfer coefficient, h_{NB} is the nucleate-boiling heat transfer coefficient, and h_{FC} is the forced-convection heat transfer coefficient.

The Foster–Zuber nucleate boiling correlation is an awkward equation, but unfortunately an integral part of the Chen calculation method.

$$h_{\mathrm{NB}} = S h_{\mathrm{FZ}} \tag{9.12}$$

where S is the suppression factor, and h_{FZ} is the nucleate-boiling heat transfer coefficient calculated from the Forster-Zuber equation. The suppression factor gradually decreases from 1 to 0 as the quality increases; the detailed definition is given later. The equation for h_{FZ} is

$$h_{\mathrm{FZ}} = \frac{0.00122 \Delta T_{\mathrm{sat}}^{0.24} \Delta p_{\mathrm{sat}}^{0.75} C_{\mathrm{p}\ell}^{0.45} \rho_{\ell}^{0.49} \kappa_{\ell}^{0.79}}{\sigma^{0.5} \lambda^{0.24} \mu_{\ell}^{0.29} \rho_{\mathrm{g}}^{0.24}} \tag{9.13}$$

This equation is dimensionally consistent, and so if SI units are used on the right-hand side the units of h_{FZ} will be $\mathrm{W/m^2 K}$. Note that $C_{\mathrm{p}\ell}$ is the liquid specific heat, κ_{ℓ} is the liquid thermal conductivity, $\Delta T_{\mathrm{sat}} = T_{\mathrm{w}} - T_{\mathrm{sat}}$, and Δp_{sat} is the difference in saturation pressure corresponding to ΔT_{sat}. Δp_{sat} is shown on the vapour pressure curve (see Fig. 9.7).

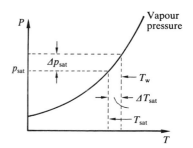

Fig. 9.7 Vapour pressure curve; explanation of the meaning of Δp_{sat}.

It will be evident that eqn (9.13) does not agree with the general finding (described in Chapter 7) about nucleate-boiling heat transfer coefficients that

$$h \propto \Delta T_{\text{sat}}^{a-1} \tag{9.14}$$

where a = 3 or 3.33. However, it must be recognized that eqn (9.13) also contains Δp_{sat}, which is a strong function of ΔT_{sat}.

For the forced convection part of the total heat transfer coefficient,

$$h_{\text{FC}} = h_\ell F \tag{9.15}$$

where h_ℓ is the single-phase liquid convective heat transfer coefficient based on the mass flow rate of liquid in the two-phase flow (not the total two-phase flow rate). This can be calculated from, for example, the Dittus–Boelter equation

$$Nu_\ell = 0.023 Re_\ell^{0.8} Pr_\ell^{0.4} \tag{9.16}$$

where

$$Nu_\ell = \frac{h_\ell d}{\kappa_\ell} \tag{9.17}$$

$$Re_\ell = \frac{G(1-x)d}{\mu_\ell} \tag{9.18}$$

$$Pr_\ell = \frac{\mu_\ell C_{p\ell}}{\kappa_\ell} \tag{9.19}$$

Chen's method mixes the nucleate-boiling heat transfer coefficient and the convective coefficient. The combination of the two is controlled by the parameters F and S.

and F is the two-phase heat transfer coefficient multiplier, which is greater than unity. Chen's special contribution was to suggest ways of calculating the factors S and F. He suggested that F should be a function of the Martinelli parameter X

$$X^2 = \frac{(dp/dz)_\ell}{(dp/dz)_g} \tag{9.20}$$

where the two pressure gradients were calculated assuming turbulent flow. $(dp/dz)_\ell$ is the frictional pressure gradient in single-phase liquid flow, where the flow rate is equal to the liquid flow rate in the two-phase flow (not the total two-phase flow rate). $(dp/dz)_g$ is defined similarly for the gas. If the friction factor for each phase is proportional to $Re^{-0.2}$, then X can be calculated simply from

$$X = \left(\frac{1-x}{x}\right)^{0.9} \left(\frac{\rho_g}{\rho_\ell}\right)^{0.5} \left(\frac{\mu_\ell}{\mu_g}\right)^{0.1} \tag{9.21}$$

The variation of F with X is shown in Fig. 9.8. Thus if the quality and the physical properties are known then F can be found.

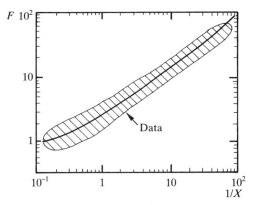

Fig. 9.8 Chen correlation for F.

In order to find the suppression factor S, the liquid Reynolds number Re_ℓ is calculated from eqn (9.18) and then a two-phase Reynolds number Re_{TP}

$$Re_{\mathrm{TP}} = Re_\ell F^{1.25} \qquad (9.22)$$

where F is the heat-transfer multiplier previously found from Fig. 9.8. S is then a function of Re_{TP} (as shown in Fig. 9.9). This then enables the total boiling heat transfer coefficient h_{B} to be calculated.

As x increases X decreases,
and $1/X$ increases,
therefore F increases,
and Re_{TP} increases,
therefore S decreases.

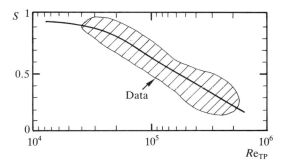

Fig. 9.9 Chen correlation for S.

As the quality goes up, the relative contribution of the nucleate boiling mechanism goes down.

9.5   Application to subcooled boiling

The Chen method is for saturated boiling. For subcooled boiling, a suitable procedure for calculating the heat flux ϕ seems to be to assume that

$$\phi = h_\ell(T_{\mathrm{w}} - T_{\mathrm{B}}) + h_{\mathrm{NB}}(T_{\mathrm{w}} - T_{\mathrm{sat}}) \qquad (9.23)$$

where T_{w} is the wall temperature and T_{B} is the bulk liquid temperature, and to calculate S on the basis that

$$Re_{TP} = Re_\ell \qquad (9.24)$$

This procedure implies that the whole of the actual temperature difference $(T_w - T_B)$ goes into the convective heat flux, but only part of it into the nucleate component. The heat flux gradually and smoothly reaches the saturated value as the bulk temperature reaches the saturation temperature.

10 Critical heat flux in flow boiling

10.1 Introduction

In Chapter 9 (see, for example, Fig. 9.2) it was seen that the heat transfer coefficient in flow boiling could fall rapidly and the wall temperature increase rapidly at some point along the heated channel. This phenomenon is known by many names, none of them entirely satisfactory.

1. Burnout. This, however, implies that the physical surface is destroyed; this does not always occur.

2. Dryout. This, however, implies a particular mechanism which does not always occur.

3. CHF—critical heat flux. This is a cumbersome term but it is widely used to refer to the phenomenon and not just the value of the heat flux at which it occurs.

4. DNB—departure from nucleate boiling. This again implies a particular mechanism which does not always occur.

5. Boiling crisis. This was widely used in the former USSR and is a very descriptive term. However, it is commonly used in the form boiling crisis of the first (or second) kind, and it is then often not clear what is meant.

CHF seems to have two distinctive characters.

1. At low quality, when it is associated with subcooled boiling or saturated nucleate boiling, it has strong similarities to pool-boiling critical heat flux, both in mechanism (formation of bubbles at the heated surface impeding liquid flow to the surface) and in behaviour (there is a hysteresis effect as the heat flux is increased and then decreased, see Fig. 10.1(a)).

2. At high quality, it is associated with convective boiling in annular flow in which the liquid film dries out and there is no hysteresis (see Fig. 10.1(b)).

10.2 Mechanisms in the subcooled region

When critical heat flux occurs in subcooled (or low quality) flow there are certain resemblances to pool-boiling critical heat flux, as already noted. A number of detailed mechanisms for this type of critical heat flux have been suggested (see Hewitt 1978).

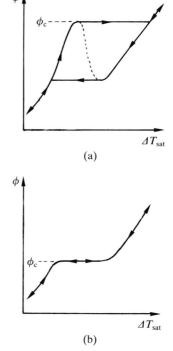

Fig. 10.1 Boiling curves for nucleate and convective boiling.

1. Near-wall bubble crowing and vapour blanketing. A kind of bubble boundary layer builds up at the heated wall. This becomes so thick and dense that it effectively stops fresh liquid reaching the surface. However, how this occurs is uncertain, and the full details of the mechanism are unclear. Is there, for example, some form of separation occurring in the 'boundary layer' which leads to stagnation?

2. Overheating at a nucleation site. It has been suggested that at very high heat fluxes, the nucleation site, which is locally dry, heats up so much during the bubble growth phase that it cannot be rewetted properly when the bubble has departed. A further discussion of the rewetting of hot surfaces is given by Whalley (1987).

3. Vapour clot or slug formation. In horizontal flow, particularly, a vapour clot or slug near a heated wall (the top wall in horizontal flow) can approach the wall very closely, so that the intermediate liquid film can evaporate; the wall is then too hot to rewet.

It seems probable that all of these mechanisms occur: the first being the usual one, the second occurring at very high heat fluxes, and the third at low flow rates (where vapour is not swept rapidly along) or in horizontal flow.

10.3   Mechanisms in the high quality region

Observations of transparent test sections and flow pattern maps show that, for most critical heat flux cases where the quality is greater than 10%, the flow pattern is annular. Many detailed studies have suggested that the critical heat flux occurs when the combined effects of entrainment, deposition, and the evaporation of the film make the film flow rate go gradually and smoothly to zero. The evidence for this is threefold.

1. Measurements of liquid film flow rate (Hewitt and Hall Taylor 1970). Experiments were conducted in a vertical heated tube in which the liquid film flow rate could be measured at the top of the heated section. This measurement was made by sucking the film off through a porous section of tube wall (see Fig. 10.2(a)). The results are shown schematically in Fig. 10.2(b). The critical heat flux point, where the heat transfer coefficient drops dramatically, occurs at the test section power which gives zero film flow rate at the outlet of the tube.

2. 'Cold patch' experiments (Bennett *et al.* 1967). Experiments were again performed in a vertical tube with steam–water flow. The tube diameter here was 8.3 mm and the pressure was 3.4 bar. Four separate test sections were used, as shown in Fig. 10.3. The results for the critical heat flux, here defined as the critical power divided by the heated area, are shown in Table 10.1.

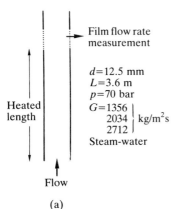

$d=12.5$ mm
$L=3.6$ m
$p=70$ bar
$G=1356$
$\quad\ 2034$ kg/m^{2}s
$\quad\ 2712$
Steam-water

(a)

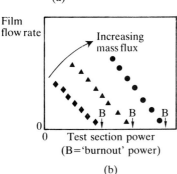

(b)

Fig. 10.2 Film flow rates near high quality critical heat flux.

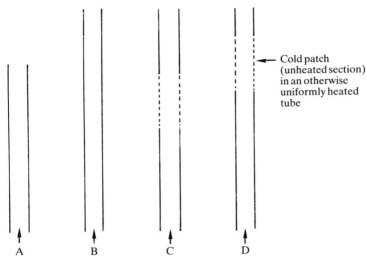

Fig. 10.3 Test sections for cold patch experiment.

Table 10.1
Results of cold patch experiments

Test section	A	B	C	D
Tube length (m)	1.8	2.4	2.4	2.4
Cold patch (unheated section) length (m)	—	—	0.6	0.6
Distance of cold patch from tube inlet (m)	—	—	1.1	1.5
Critical heat flux ϕ_c (kW/m^2)	652	400	617	690

'A', 'C', and 'D' all have a heated length of 1.8 m and so are directly comparable, but the position of the cold patch can apparently lead to an increase or a decrease in the critical heat flux. The explanation lies with the details of the annular flow and the variation of the entrained liquid flow with the tube length. Figure 10.4 shows the experimentally determined variation of entrained liquid flow with quality in the four test sections, and also the equilibrium entrained liquid flow rate. CHF will occur when the line for each tube meets the total liquid line, because then all the liquid is entrained and the film flow is zero.

Heating the flow increases the quality and the point representing the flow conditions moves along the curve to the right. The system also tries to reach the hydrodynamic equilibrium line. Thus when the operating line crosses the equilibrium line it does so with zero gradient. Over the length of a cold patch the quality does not change, so the point can move vertically—again towards the equi-librium line. Test section 'C' has the cold patch near the entrance, so it is encountered at low quality. The liquid entrainment rises over the length of this cold patch and the critical heat flux is lower than for a uniformly heated tube with the same heated length (test section 'A'). Test section 'D', on the other hand, has the cold

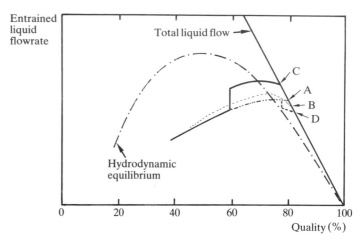

Fig. 10.4 Entrainment diagram for cold patch experiment.

patch nearer the exit, so it is encountered at higher quality. The entrainment falls over the length of this cold patch and the critical heat flux is higher than for the uniformly heated tube. This is the reason why the cold patch can increase or decrease the critical heat flux.

3. Physically based models of the flow. The third piece of evidence for the film dryout mechanism is the relative success of models based on it for predicting actual values of critical heat flux, see Whalley (1987).

10.4   Parametric trends and forms of correlation

The main parametric trends for the critical heat flux for a uniformly heated tube are illustrated in Fig. 10.5.

1. Figure 10.5(a). The critical heat flux usually varies linearly with the inlet subcooling Δh_s (J/kg). The inlet subcooling is the liquid saturation enthalpy minus the actual liquid enthalpy at the test section entrance.

2. Figure 10.5(b and c). The critical heat flux increases with mass flux and tube diameter.

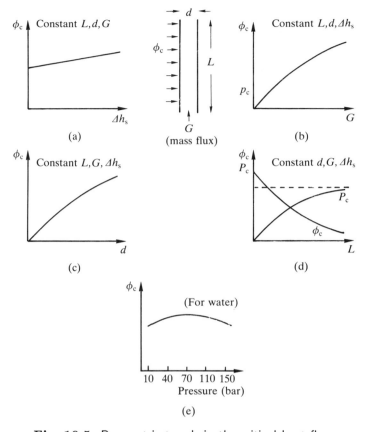

Fig. 10.5 Parametric trends in the critical heat flux.

3. Figure 10.5(d). The critical heat flux asymptotes to zero as the tube length increases. The critical power P_c, which is given by

$$P_c = \pi d L \phi_c \qquad (10.1)$$

asymptotes to a constant power sufficient to vaporize all the liquid and thus produce a quality of unity.

4. Figure 10.5(e). For steam–water flows the critical heat flux goes through a maximum at about 70 bar. From Chapter 8 it can be seen that the pool-boiling critical heat flux for steam–water also passes through a maximum at about 70 bar. The connection between these facts is not clear.

The common forms of correlation for critical heat flux in flow boiling can be expressed at two graphical plots, as in Fig. 10.6. In this figure x_c is quality at which the critical condition occurs, and L_B is the length over which boiling occurs (m).

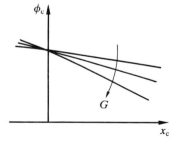

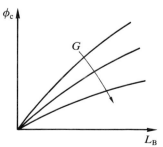

Fig. 10.6 Forms of critical heat flux correlation.

Figure 10.6(a) suggests that CHF is a purely local condition: if the local values of x and ϕ_c correspond to the critical values, then CHF will occur.

Figure 10.6(b) suggests that CHF is an integral process: it contains the boiling length which encompasses some information about the history of the flow.

However, most experiments are performed in uniformly heated tubes. A heat balance over the boiling length L_B gives

$$\pi d L_B \phi_c = \frac{\pi d^2}{4} G \lambda x_c \tag{10.2}$$

or

$$L_B \phi_c = \frac{dG\lambda}{4} x_c \tag{10.3}$$

and so a correlation of ϕ_c against x_c contains no less information than one of ϕ_c against L_B. The most usual form of correlation encountered in practice is one of ϕ_c against x_c, as in Fig. 10.6(a). This type of correlation is known as a 'local conditions' correlation. The critical heat flux is assumed to be a function of the local conditions only.

10.5   Local conditions correlations

A typical correlation of this type is that due to MacBeth (1963). A simplified version is

$$\phi_c = A\lambda G^{\frac{1}{2}}(1-x) \tag{10.4}$$

This was developed for water, and the constant A is approximately $0.25 \text{ kg}^{\frac{1}{2}}/\text{ms}^{\frac{1}{2}}$. On its own it is of little use as it relates two unknown quantities ϕ_c and x. It can be combined with a heat balance to eliminate either ϕ_c or, more commonly, the quality x. It can, however, be used to decide where in the tube the critical condition will occur. Both uniformly heated tubes and non-uniformly heated tubes are considered.

1. Uniformly heated tubes (see Fig. 10.7). Figure 10.7(a) shows the basic local conditions correlation eqn (10.4). Figure 10.7(b) shows the variation of quality with length for three different heat fluxes. The result of eliminating the quality is shown in Fig. 10.7(c). The critical heat flux can be compared with the actual heat flux shown in Fig. 10.7(d). The comparison is shown directly in Fig. 10.7(e). At heat flux levels '1' and '2' the two lines do not touch or intersect, so CHF does not occur. Raising the heat flux to level '3' causes CHF to occur for the first time at the exit of the tube. This is a general result: in uniformly heated vertical tubes, CHF always occurs first at the tube outlet.

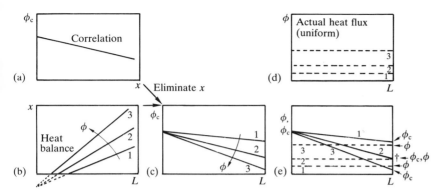

Fig. 10.7 Position of the CHF for a uniformly heated tube. CHF first occurs at the end of the tube.

2. Non-uniform heating. In this case a sinusoidal variation of heat flux along the tube is assumed. The result of repeating the previous steps is shown in Fig. 10.8. In Fig. 10.8(e) at the least flux level, level '1', the actual heat flux and the critical heat flux just touch, so CHF occurs. It occurs some distance upstream from the end of the tube. As the heat flux is increased to level '2' the CHF area spreads upstream and downstream. Whenever the heat flux falls with distance along the test section, there is the possibility that CHF may first occur at a point before the tube outlet.

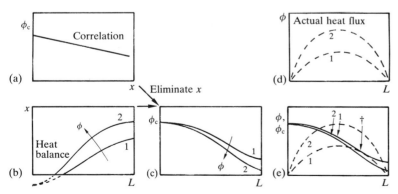

Fig. 10.8 Position of the CHF for a non-uniformly heated tube. CHF first occurs upstream of the end of the tube.

10.6   Calculation of critical heat flux

Due to the vast amount of data collected for steam–water flow in vertical tubes, it is possible to recommend calculation methods for this situation.

1. Bowring (1972) correlation. This has the form

$$\phi_{\mathrm{c}} = \frac{A + B\Delta h_{\mathrm{s}}}{C + L} \tag{10.5}$$

   where A, B, and C are function of p, λ, d, and G. Note that the Bowring correlation can be rearranged into the local-conditions form, where the critical heat flux is a function of the local quality. The details of the Bowring correlation are given by Whalley (1987).

2. Groeneveld (1982) tabular method. Groeneveld has developed a method which gives tables for the function

$$\phi_{\mathrm{c}} = f(p, G, x) \tag{10.6}$$

   There are correction factors for tube diameter d and tube length L. Note that this, again, is basically a kind of local conditions form of expression. Groeneveld's tabular method is a development of earlier tables published by Russian workers (see Doroshchuk *et al.* 1975; Collier and Thome 1994).

For fluids other than steam–water, there are two options available.

1. Ahmad (1973) scaling rules. Of the very many possible dimensionless groups for characterizing critical heat flux, Ahmad identified the important ones as

$$\frac{L}{d}, \frac{\rho_\ell}{\rho_{\mathrm{g}}}, \frac{\Delta h_{\mathrm{s}}}{\lambda}, \psi, \quad \text{and} \quad \frac{\phi_{\mathrm{c}}}{G\lambda}$$

   Here

$$\psi = \frac{Gd}{\mu_\ell} \left(\frac{\mu_\ell^2}{\sigma d\rho_\ell} \right)^{\frac{2}{3}} \left(\frac{\mu_{\mathrm{g}}}{\mu_\ell} \right)^{\frac{1}{5}} \tag{10.7}$$

   If the first four dimensionless groups are the same in the two different fluid systems, then the fifth dimensionless group will also be the same in the two systems. Hence to calculate ϕ_{c} for a fluid X from information about critical heat flux in water, the following procedure is recommended.

(a) Make L/d the same for the water system and the fluid X system. Normally, the two systems have equal lengths and equal diameters.

(b) Make ρ_ℓ/ρ_g the same for the water system and the fluid X system. This determines the water pressure.

(c) Make $\Delta h_s/\lambda$ the same for the water system and the fluid X system. This determines the water inlet enthalpy.

(d) Make ψ the same for the water system and the fluid X system. This determines the water mass flux.

(e) Calculate ϕ_c, for the now fully determined water conditions. This can be done by, for example, the Bowring correlation described above.

(f) Hence find ϕ_c for the fluid X, knowing that

$$\left(\frac{\phi_c}{G\lambda}\right)_{water} = \left(\frac{\phi_c}{G\lambda}\right)_X \qquad (10.8)$$

2. Katto generalized correlation (see Katto and Ohne 1984, referred to here simply as 'Katto'). In a long series of papers Katto has assumed that

$$\phi_c = XG(\lambda + K\Delta h_s) \qquad (10.9)$$

and then correlated X and K. Note, therefore, that he has assumed that the critical heat flux varies linearly with the inlet subcooling. This is usually a good assumption. He further assumed that X and K were functions of three dimensionless groups

$$X, K = f\left(\frac{L}{d}, \frac{\rho_\ell}{\rho_g}, \frac{\sigma\rho_\ell}{G^2L}\right) \qquad (10.10)$$

The basic Katto equation (eqn (10.9)) can be rearranged to give

$$\frac{\phi_c}{G\lambda} = X + XK\frac{\Delta h_s}{\lambda} \qquad (10.11)$$

It is now evident that the Ahmad scaling rules and the Katto form are very similar. Of Ahmad's five dimensionless groups, Katto uses four. The disagreement is about the fifth:

Ahmad	Katto
$\frac{Gd}{\mu_\ell}\left(\frac{\mu_\ell^2}{\sigma d\rho_\ell}\right)^{\frac{2}{3}}\left(\frac{\mu_g}{\mu_\ell}\right)^{\frac{1}{5}}$	$\frac{\sigma\rho_\ell}{G^2L}$

The Katto correlation has probably been better tested that the Ahmad scaling rules. Katto used test data for:

water	ammonia	benzene	ethanol
helium	hydrogen	nitrogen	potassium
R12	R21	R22	R113

He seemed to obtain good results for most data. The Katto correlation is given in detail by Whalley (1987).

11 Condensation

11.1 Modes of condensation

Condensation can occur in two ways, as illustrated in Fig. 11.1. First there could be filmwise condensation, as in Fig. 11.1(a) where the condensate (that is the liquid formed when the vapour condenses) forms a continuous film on the cold solid surface. The latent heat released on condensation is conducted through this film from the interface where the condensation occurs and is removed through the wall. The other mechanism is dropwise condensation, shown in Fig. 11.1(b). In this case the condensate forms in drops which do not wet the solid surface well. The drops therefore do not form a continuous film; instead they reach a certain critical size and then run off the surface. The surface is left dry and another drop can begin to form, rather like the formation of bubbles in nucleate boiling from a nucleation site. Because in dropwise condensation the vapour is in direct contact with the solid surface, the condensation heat transfer coefficients are significantly larger than in filmwise condensation. However, dropwise condensation is difficult to promote reliably. Special surface coatings do give drop formation, but the effectiveness of such surface treatments decreases with time. The normal design practice is to assume that filmwise condensation takes place.

For a general review of condensation see Griffith and Butterworth (1982).

11.2 Filmwise condensation on a vertical surface

For a laminar falling film the calculation of the heat transfer coefficient can be split into a number of parts.

1. Hydrodynamics. As was done for flooding flow in Chapter 6, consider a force balance on the outer part of the film (see Fig. 11.2). This time, however, the gas density is taken into account. For steady flow, in which the streamlines in the liquid are vertical

$$\delta p = \rho_g g \delta z \qquad (11.1)$$

and so a force balance in the vertical direction gives

$$\tau \delta z + \delta p(\delta - y) = \rho_\ell g(\delta - y)\delta z \qquad (11.2)$$

or simplifying using eqn (11.1)

$$\tau = (\rho_\ell - \rho_g)g(\delta - y) \qquad (11.3)$$

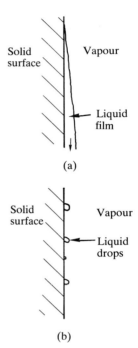

Fig. 11.1 Modes of condensation: (a) filmwise condensation, and (b) dropwise condensation.

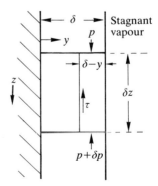

Fig. 11.2 Control volume for force balance in falling film flow.

Then, putting the laminar flow value for the shear stress

$$\tau = \mu_\ell \frac{\mathrm{d}u}{\mathrm{d}y} \tag{11.4}$$

and so, using the boundary condition that when $y = 0$, $u = 0$

$$u = \frac{(\rho_\ell - \rho_\mathrm{g})gy}{\mu_\ell} \left(\delta - \frac{y}{2}\right) \tag{11.5}$$

Note that we have also put (implicitly) that the interfacial shear stress is zero, that is that $\tau = 0$ when $y = \delta$. The mass flow rate M per unit width of film is then given by

$$M = \int_0^\delta \rho_\ell u \, \mathrm{d}y = \frac{\rho_\ell(\rho_\ell - \rho_\mathrm{g})g}{\mu_\ell}\frac{\delta^3}{3} \tag{11.6}$$

and differentiating, the change in the mass flow rate can be related to the changes in the film thickness

$$\frac{\mathrm{d}M}{\mathrm{d}z} = \frac{\rho_\ell(\rho_\ell - \rho_\mathrm{g})g}{\mu_\ell}\delta^2\frac{\mathrm{d}\delta}{\mathrm{d}z} \tag{11.7}$$

Equation (11.7) is then valid as long as the film thickness changes slowly so that the streamlines are nearly vertical. This is true as long as $d\delta/dz \ll 1$.

2. Heat transfer. As the vapour condenses, its specific enthalpy changes by λ the latent heat. Strictly, the change is slightly larger since the condensate becomes subcooled in the liquid film. However, to a good approximation

$$\phi = \lambda\frac{\mathrm{d}M}{\mathrm{d}z} \tag{11.8}$$

Another equation for the heat flux comes from the fact that in laminar flow the heat is conducted through the film: there is no convection. The temperature profile is linear, so from the conduction equation

$$\phi = +\frac{\kappa_\ell(T_\mathrm{w} - T_\mathrm{sat})}{\delta} = +\frac{\kappa_\ell}{\delta}\Delta T_\mathrm{sat} \tag{11.9}$$

where, in this equation, the positive sign arises because:

at $y = 0$, $T = T_\mathrm{w}$, and
at $y = \delta$, $T = T_\mathrm{sat}$.

Equating eqns (11.8) amd (11.9) for the heat flux

$$\frac{dM}{dz} = \frac{\kappa_\ell \Delta T_{\text{sat}}}{\lambda \delta} \tag{11.10}$$

3. Solution for heat transfer coefficient. Now we have two equations for dM/dz; one from the hydrodynamics (eqn (11.7) and one from the heat transfer (eqn (11.10)). Equating the two values of dM/dz, and rearranging gives

$$\delta^3 d\delta = \frac{\mu_\ell \kappa_\ell \Delta T_{\text{sat}}}{\rho_\ell(\rho_\ell - \rho_{\text{g}})g\lambda} dz \tag{11.11}$$

Then integrating, and assuming that the film thickness is zero at $z = 0$, eqn (11.11) gives

$$\delta = \left[\frac{4\mu_\ell \kappa_\ell \Delta T_{\text{sat}} z}{\rho_\ell(\rho_\ell - \rho_{\text{g}})g\lambda}\right]^{\frac{1}{4}} \tag{11.12}$$

where δ is the film thickness at a distance z from the top of the plate. The heat transfer coefficient h_z at this distance z is given by

$$h_z = \frac{\kappa_\ell}{\delta} \tag{11.13}$$

So using eqn (11.12),

$$h_z = \left[\frac{\rho_\ell(\rho_\ell - \rho_{\text{g}})g\lambda\kappa_\ell^3}{4\mu_\ell \Delta T_{\text{sat}} z}\right]^{\frac{1}{4}} \tag{11.14}$$

However, we are usually interested, not in the local heat transfer coefficient h_z, but the average value h over the whole plate. This value can be obtained by averaging h_z

$$h = \frac{1}{L}\int_0^L h_z dz \tag{11.15}$$

The result of this integration is that the average condensation heat transfer coefficient h is given by

$$h = \frac{2\sqrt{2}}{3}\left[\frac{\rho_\ell(\rho_\ell - \rho_{\text{g}})g\lambda\kappa_\ell^3}{\mu_\ell \Delta T_{\text{sat}} L}\right]^{\frac{1}{4}} \tag{11.16}$$

This equation was first obtained by Nusselt in 1916. The effect of the subcooling of the liquid in the film is easily taken into account if it is assumed that there is a linear temperature profile in the film. The result is that the latent heat in eqn (11.16) should be replaced by a modified latent heat λ'

$$\lambda' = \lambda + \frac{3}{8}C_{p\ell}(T_{\text{sat}} - T_{\text{w}}) \tag{11.17}$$

Rohsenow (1956) pointed out that because the condensation is occurring all the time, the film never has a chance to adopt the linear temperature profile. Because of this effect, the factor of $\frac{3}{8}$ in eqn (11.17) should be modified to 0.68.

As an example, calculate the average heat transfer coefficient of condensation for steam at 1 bar on a vertical surface 1 m long when the surface temperature is 60 °C. The physical properties are:

$$
\begin{aligned}
\rho_\ell &= 1000 \text{ kg/m}^3 \\
\lambda &= 2256 \times 10^3 \text{ J/kg} \\
\rho_{\text{g}} &= 0 \\
\kappa_\ell &= 0.68 \text{ W/mK} \\
\mu_\ell &= 0.283 \times 10^{-3} \text{ Ns/m}^2
\end{aligned}
$$

The remaining variables required are then $\Delta T_{\text{sat}} = 100 - 60 = 40\,°\text{C}$, and $L = 1$ m, so from eqn (11.16) The remaining variables required are then $\Delta T_{\text{sat}} = 60\,°\text{C}$, and $L = 1$ m, so from eqn (11.16)

$$h = \frac{8^{\frac{1}{2}}}{3}\left(\frac{10^3 \times 10^3 \times 9.81 \times 2256 \times 10^3 \times 0.68^3}{0.283 \times 10^{-3} \times 40 \times 1}\right)^{\frac{1}{4}} = 4700 \text{ W/m}^2\text{K}$$

This is a typical value for the condensation heat transfer coefficient.

11.3   Filmwise condensation on a horizontal tube

For condensation on the outside of a horizontal tube, the condensate film drains down the tube and drips off the bottom as shown in Fig. 11.3. The derivation of the equation for the average condensation heat transfer coefficient is similar to, but more complicated than, the derivation of eqn (11.16). The final result is that for the horizontal tube

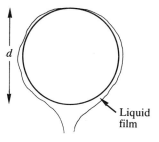

Fig. 11.3 Filmwise condensation on the outside of a horizontal tube.

$$h = 0.73\left[\frac{\rho_\ell(\rho_\ell - \rho_{\text{g}})g\lambda\kappa_\ell^3}{\mu_\ell\Delta T_{\text{sat}}d}\right]^{\frac{1}{4}} \tag{11.18}$$

The equation is in exactly the same form as eqn (11.16), except that the constant in the equation for a flat place was $8^{\frac{1}{2}}/3 = 0.94$. The present value is lower because on a tube the film tends to be thicker. Typical heat transfer coefficients on a horizontal tube are larger than for a flat plate, so repeating the example in the previous section with $d = 0.02$ m instead of $L = 1$ m, gives $h = 9700 \text{ W/m}^2\text{K}$ (about twice the previous value). For this reason condenser tubes are usually arranged to be horizontal rather than vertical. However, we usually have bundles of horizontal tubes not just one, as in Fig. 11.4. The condensate from one tube drips on to the next one below. The effect is almost as if the tube diameter was larger than the actual value. In the vertical plate derivation this would mean that the film thickness was not equal to zero at $z = 0$ for the second and subsequent plate in a vertical line. So, if there are n tubes in a vertical line, then the average condensation heat transfer coefficient h_n over the n tubes is given by

$$h_\mathrm{n} = \frac{h}{n^{\frac{1}{4}}} \qquad (11.19)$$

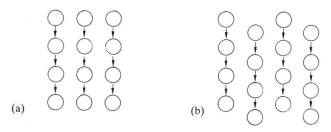

Fig. 11.4 Drainage of condensate in a tube bundle: (a) in-line array, and (b) staggered array.

So, if $h = 9700$ W/m^{2}K (as above), and $n = 18$, then

$$h_\mathrm{n} = \frac{9700}{18^{\frac{1}{4}}} = 4700 \text{ W/m}^2\text{K}$$

which is approximately the same result as for the vertical plate of length 1 m.

Actually, the downward flow pattern of the condensate is more favourable than envisaged in the simple extension to the Nusselt analysis, as illustrated in Fig. 11.5. Figure 11.5(a) shows the idealized flow, where the liquid flows off one tube in a continuous sheet. However, this is not what actually happens: as shown in Fig. 11.5(b), the liquid drips off at discrete points. The separation of the drip points is roughly given by the Taylor wavelength

$$\lambda_\mathrm{T} = 2\pi\sqrt{3}\left[\frac{\sigma}{(\rho_\ell - \rho_\mathrm{g})g}\right]^{\frac{1}{2}} \qquad (11.20)$$

as in Chapter 8 on pool-boiling critical heat flux. This alteration to the ideal flow means that the $n^{-\frac{1}{4}}$ factor in eqn (11.19) is not correct. Experimentally, a better version of that equation is (see Collier and Thome 1994).

$$h_\mathrm{n} = \frac{h}{n^{\frac{1}{6}}} \qquad (11.21)$$

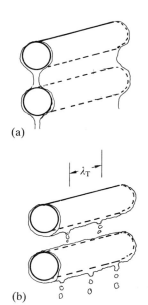

Fig. 11.5 (a) Idealized 'Nusselt' condensation and (b) actual flow of condensate from a horizontal tube to a tube below.

So, taking the figures used before, $h = 9700$ W/m^{2}K and $n = 18$, then

$$h_{\mathrm{n}} = 6000 \text{ W/m}^2\text{K}$$

Real bundles are usually arranged so that:

1. the tubes are staggered as in Fig. 11.4(b), not in a vertical line as in Fig. 11.4(a) (this means that the condensate from a tube at the top of the bundle flows over as few other tubes as possible); and

2. in very large bundles the condensate is removed at intervals down the bundle (so that n in eqn (11.21) does not get too large).

11.4   Real condensation

Real condensation differs from the above analysis (the Nusselt analysis) in a number of ways.

1. The film is almost never smooth. Ripples on the surface increase the surface area and stir up the film. These effects can increase the heat transfer coefficient by about 20%.

2. There is a shear force exerted by the flowing vapour on the liquid film. Normally, this is arranged to make the film thinner and so to increase the heat transfer coefficient.

3. The film may become turbulent. This happens at values of the film Reynolds number ($4M/\mu_\ell$) in the range 1000 to 2000. Again, turbulence increases the heat transfer coefficient as compared to the laminar flow values. The effect of the turbulence is to increase the effective viscosity of the liquid: this makes the liquid film thicker. This effect is, however, more than counteracted by the increased heat transfer due to convection effects in the liquid.

4. Multi-component effects. The simplest of these is the effect of an 'incondensable' gas (such as, for example, air) on the condensation of steam. Such effects are discussed in some detail in later. However, it can be noted here that in many cases the presence of an incondensable gas can give rise to condensation heat transfer coefficients which are far less than expected.

In contrast to the large effects of an incondensable gas, the first three effects all lead to fairly modest increases in the condensation heat transfer coefficient.

11.5 Condensation inside a horizontal tube

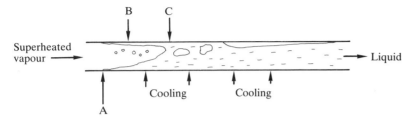

Fig. 11.6 Total condensation of a pure vapour inside a horizontal tube.

Condensation inside a horizontal tube is a common industrial situation. The vapour may enter the tube superheated, so the flow will be as shown in Fig. 11.6. At point 'A' the wall is first wet with condensate, the liquid film on the walls of the tube is slightly subcooled and the vapour is still superheated. The region before 'A' is known as the dry wall desuperheating region; the region after 'A' is known as the wet wall desuperheating region. At point 'B' liquid drops are torn off the liquid film into the super-heated vapour where they then evaporate. From 'A' until 'C' the flow is an annular flow. At first, near 'A', there is a significant degree of thermodynamic non-equilibrium (superheated vapour and subcooled liquid). The region from 'A' to 'C' is a possible region for the application of annular flow modelling, see Whalley (1987). From point 'C' onwards the flow goes through various patterns until the last vapour condenses (in a total condenser) to give, finally, subcooled liquid.

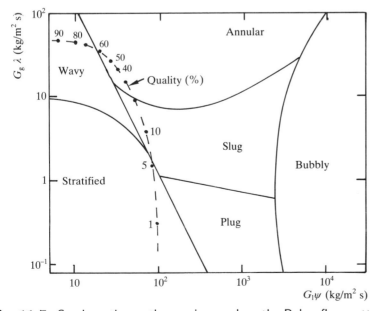

Fig. 11.7 Condensation path superimposed on the Baker flow pattern chart.

In analysing the condensation, it is interesting to note the flow patterns which are found. The path of the condensation can be plotted on a flow regime map; in Fig. 11.7 a typical condensation path is plotted on a Baker flow pattern map (see Chapter 2), where λ and ψ are defined. As in boiling flow, a substantial length of the tube is in the annular flow region. The condensation coefficient has been analysed for various flow patterns, see Collier and Thome (1994) and Owen and Lee (1983). See these references also for details of the many more specialised methods which have been suggested for use in particular flow patterns.

11.6   Condensation of a vapour and an incondensable gas

In a case such as the condensation of steam which contains some air, there are concentration gradients of the air and the steam near the interface as shown in Fig. 11.8.

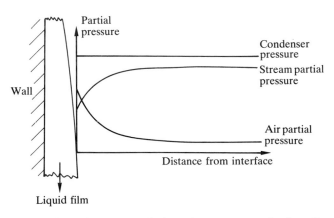

Fig. 11.8 Partial pressure of air and steam near the interface in a condenser.

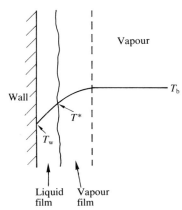

Fig. 11.9 Variation of temperature near and in the liquid film during the condensation of a vapour containing an incondensable gas.

Although in this case the partial pressure of the air in the condenser is quite small, the partial pressure at the interface is quite substantial. The air is diffusing away from the interface down its concentration gradient, and being pulled towards the interface by the flow of steam. The air partial pressure profile is the result of these competing effects. It is evident that a proper calculation of the condensation rate would involve a complex boundary layer calculation. Practical calculation of condensation coefficients has relied on a number of drastically simplified theories, such as that of Silver (1947) and Bell and Ghaly (1973). Plotting the temperature distribution near the wall (see Fig. 11.9), the whole temperature difference, $T_b - T_w$ can be divided into two parts

$$T_b - T_w = (T_b - T^*) + (T^* - T_w) \qquad (11.22)$$

Assuming the heat flux is the same across the interface as across the wall then, since $\Delta T = \phi/h$, eqn (11.22) can be rewritten as

$$\frac{\phi_w}{h} = \frac{\phi_g}{h_g} + \frac{\phi_w}{h_f} \qquad (11.23)$$

where ϕ_w is the wall heat flux (W/m^2), ϕ_g is the 'sensible' heat flux (W/m^2)—this is the heat flux which arises because of the temperature differences in the gas phase—h is the condensation heat transfer coefficient (W/m^2K), h_f is the heat transfer coefficient across the liquid film (W/m^2K) (this could be calculated by means of eqn (11.16)), and h_g is the single-phase gas heat transfer coefficient (W/m^2K).

The energy flow from the vapour core to the interface is made up of two parts:

1. that which is released as latent heat by condensation at the interface; and

2. that which is represented by cooling of the vapour as it approaches the interface: this is the sensible heat.

It is assumed that the sensible heat resistance is at least as great as the liquid film resistance.

We can now write

$$\frac{1}{h} = \frac{1}{h_f} + \frac{\phi_g/\phi_w}{h_g} \qquad (11.24)$$

In order to produce a solution, we need a value for ϕ_g/ϕ_w, which can be estimated by arguing that

$$\phi_g = \dot{m} C_{pg} \Delta T x \qquad (11.25)$$

where $\dot{m}$ is the total mass flow towards the surface per unit area of wall (kg/m^2s), C_{pg} is the gas phase specific heat, and x is the quality of this mass flow. Then the total heat flux ϕ_w is given by

$$\phi_w = \dot{m} \Delta h \qquad (11.26)$$

Eliminating $\dot{m}$ between eqns (11.25) and (11.26) we find that

$$\frac{\phi_g}{\phi_w} = x C_{pg} \frac{dT}{dh} \qquad (11.27)$$

where x is the local quality, and dT/dh is the slope of the condensation curve. The condensation curve is a graph of the enthalpy released against temperature as the condensation proceeds (see Fig. 11.10). The enthalpy is the total enthalpy at any point: the sum of the enthalpies of the liquid phase, of the condensable vapour, and of the non-condensable gas.

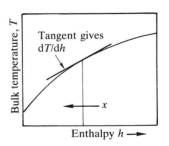

Fig. 11.10 Condensation curve.

As an example, consider the calculation of a mixture originally of:

steam flow rate   0.9   kg/s,
air flow rate   0.1   kg/s, and at
total pressure p 1   bar.

The initial partial pressures are calculated from the ratio of the numbers of moles flowing

$$p_{steam} = \frac{\frac{0.9}{18}}{\frac{0.9}{18} + \frac{0.1}{29}} \times 1 = 0.935 \text{ bar}$$

$$p_{air} = p - p_{steam} = 1 - 0.935 = 0.065 \text{ bar}$$

Here 18 and 29 are the molecular weights of steam and air. The condensation will begin when the temperature is the steam saturation temperature equivalent to 0.935 bar, which is 97.6 °C. Then, choosing temperatures below this, the condensation curve can be constructed. For example, take $T = 90$ °C. At this temperature

$$p_{steam} = p_{sat} = 0.701 \text{ bar}$$

and so

$$p_{air} = p - p_{steam} = 1 - 0.701 = 0.299 \text{ bar}$$

The volume occupied by 0.1 kg of air (the mass flowing in 1 s) at this temperature $(273 + 90 = 363 \text{ K})$ and pressure $(0.299 \times 10^5 \text{ N/m}^2)$ can be found from the ideal gas equation

$$V = \frac{m_{air} RT}{p_{air}} = \frac{0.1 \times \frac{8314}{29} \times 363}{0.299 \times 10^5} = 0.348 \text{ m}^3$$

From steam tables, the specific volume of saturated steam at 90 °C is 2.361 m³/kg, so the mass of steam associated with 0.1 kg of air is

$$\frac{0.348}{2.361} = 0.147 \text{ kg}$$

Hence $0.9 - 0.147 = 0.753$ kg of steam have condensed. The enthalpy of the mass flowing in 1 s can now be calculated. The three bracketed terms below are the enthalpies of the air, the liquid water, and the steam.

$$h = (0.1 \times 1.005 \times 90) + (0.753 \times 376.9) + (0.147 \times 2660.1) = 683.9 \text{ kJ/kg}$$

and

$$x = \frac{0.1 + 0.147}{1} = 0.247$$

Note that just before condensation began, $T = 97.6 \,^\circ\text{C}$, $x = 1$, and

$$h = (0.1 \times 1.005 \times 97.6) + (0.9 \times 2669.0) = 2411.9 \text{ kJ/kg}$$

Taking other temperatures, the condensation curve can be calculated: the result is shown in Fig. 11.11.

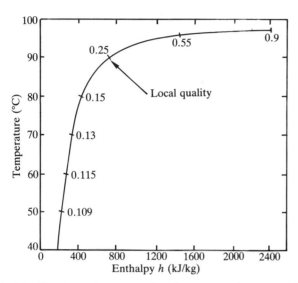

Fig. 11.11 Condensation curve for air–steam mixture used in the example.

Now since, from eqns (11.23) and (11.27)

$$\frac{1}{h} = \frac{1}{h_\text{f}} + \frac{xC_\text{pg}\mathrm{d}T/\mathrm{d}h}{h_\text{g}} \tag{11.28}$$

h falls rapidly as the quality falls. This is because $\mathrm{d}T/\mathrm{d}h$ becomes very large and so $1/h$ is large and therefore h is small. This means that the first part of the condensation can be performed without difficulty, but that it is very difficult (that is, a large surface area is required) to condense a very large fraction of the steam. Of course, the shape of the condensation curve alters as the inlet flows of air and steam are altered (or more specifically, as their ratio is altered). Reduction of the overall heat transfer coefficient is not so severe if the initial fraction of air in the steam, here assumed to be 10% by weight, is lower.

However, turbine exhaust condensers in a steam power station always have pumps to remove the air which inevitably leaks into the system. They are arranged so as to remove as much air and as little steam as possible (as the steam lost represents a loss of water which is chemically treated in order to inhibit corrosion, and also to minimize the pump power). One possible arrangement is shown in Fig. 11.12.

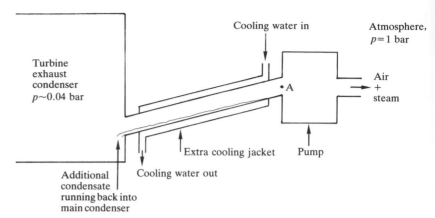

Fig. 11.12 Venting arrangement in a turbine exhaust condenser.

The object is to reduce the steam partial pressure at 'A' to as low a value as possible. This is dependent on the temperature of the cooling water: obviously the lower the temperature of the cooling water the better.

References

AHMAD, S.Y. (1973). Fluid to fluid modelling of critical heat flux, a compensated distortion model. *Int. J. Heat Mass Transfer* **16** 641–61.

ALEKSEEV, V.P., POBEREZKIN, A.E., and GERASIMOV, P.V. (1972). Determination of flooding rates in regular packings. *Heat Transfer Soviet Research* **4** (No. 6), 159–63.

ANDEEN, G.B. and GRIFFITH, P. (1968). Momentum flux in two-phase flow. *J. Heat Transfer* **90** 211–22.

BAKER, O. (1954). Simultaneous flow of oil and gas. *Oil Gas J.* **53** 185.

BANERJEE, S., RHODES, E., and SCOTT, D.S. (1969). Studies on cocurrent gas–liquid flow in helically coiled tubes: flow patterns, pressure drop, and hold-up. *Can. J. Chem. Engnrs* **47** 445–517.

BARNEA, D., SHOHAN, O., TAITEL, Y., and DUKLER, A.E. (1980). Flow pattern transition for horizontal and inclined pipes: experimental and comparison with theory. *Int.J. Multiphase Flow* **6** 217–25.

BAROCZY, C.J. (1966). A systematic correlation for two-phase pressure drop. *Chem. Eng. Prog. Symp. Ser.* **62** (no. 64) 232–49.

BEATTIE, D.R.H. and WHALLEY,P.B. (1981). A simple two-phase frictional pressure drop calculation method. *Int. J. Multiphase Flow* **8** 83–7.

BELL, K.J. and GHALY, M.A. (1973). An approximate generalised design method for multi-component/partial condensers. *AIChE Symp. Ser* **69** 72–9.

BELLMAN, R. and PENNINGTON, R.H. (1954). Effects of surface tension and viscosity on Taylor instability. *Quart. J. Appl. Math.* **12** 151.

BENNETT, A.W., HEWITT, G.F., KEARSEY, H.A., and KEEYS, R.K.F. (1967). Heat transfer to steam-water mixtures flowing in uniformly heated tubes in which the critical heat flux has been exceeded. *AERE-R5373*.

BENNETT, A.W., HEWITT, G.F., KEARSEY, H.A., KEEYS, R.K.F., and PULLING, D.J. (1967). Studies of burnout in boiling heat transfer to water in round tubes with non-uniform heating. *Trans. Instn Chem. Engnrs* **45** 319–33.

BERGLES, A.E. (1969). Two-phase flow structure observations for high pressure water in a rod bundle. *Proc. ASME Symp. Two–Phase Flow Heat Transfer in Rod Bundles*, 47–55.

BERGLES, A.E. and ROHSENOW, W.M. (1964). the determination of forced-convection surface boiling heat transfer. *J. Heat Transfer* **86** 365–82.

BLANDER, R.M. and KATZ, J.L. (1975). Bubble nucleation in liquids. *AIChE J* **21** 833–43.

BOWRING, R.W. (1972). A simple but accurate round tube uniform heat flux dryout correlation over the pressure range 0.7–17 MN/m^2. *AEEW-R789*.

BRYCE, W.M. (1977). A new flow-dependent slip correlation which gives hyperbolic steam–water mixture equations. *AEEW - R1099*.

CHEN, J.C. (1963). A correlation for boiling heat transfer to saturated fluids in convective flow. *ASME paper 63-HT-34*. Presented at 6th National Heat Transfer Conference, Boston.

CHISHOLM, D. (1967). A theoretical basis for the Lockhart-Martinelli correlation for two-phase flow. *Int. J. Heat Mass Transfer* **10** 1767–78.

CHISHOLM, D. (1972). An equation for velocity ratio in two-phase flow. *NEL Report 535*.

CHISHOLM, D. (1973). Pressure gradients due to friction during the flow of evaporating two- phase mixtures in smooth tubes and channels. *Int. J. Heat Mass Transfer* **16** 347–58.

CHISHOLM, D. (1978). Influence of pipe surface roughness on friction pressure gradient during two-phase flow. *J. Mech. Eng. Sc.* **20** 353–4.

CHISHOLM, D. (1983). *Two-phase flow in pipelines and heat exchangers*. George Godwin, London.

COLLIER, J. G. and THOME, J. R. (1994). *Convective boiling and condensation*. (3rd edition). Oxford University Press.

COOPER, M.G. (1984). Saturated nucleate pool boiling—a simple correlation. *1st UK National Heat Transfer Conference* (I. Chem. E. Symp. series No. 86) **2** 785–93.

DAVIS, E.J. and ANDERSON, G.H. (1966). The incipience of nucleate boiling in forced convection flow. *AIChE J.* **12** 774–80.

DOROSHCHUK, V.E., LEVITAN, L.L., and LANTZMAN, F.P. (1975). Investigations into burnout in uniformly heated tubes. *ASME paper 75-WA/HT-22.*

DUKLER, A.E., WICKS, M., and CLEVELAND, R.G. (1964). Frictional pressure drops in two-phase flow. *AIChE J* **10** 44.

GOLAN, P.L. and STENNING, A.H. (1969). Two-phase vertical flow maps. *Proc.Instn.Mech.Engnrs* **184(3C)** 110–16.

GRIFFITH, P. and BUTTERWORTH, D. (1982). Condensation. In Hetsroni (1982).

FRIEDEL, L. (1979). Improved friction pressure drop correlations for horizontal and vertical two-phase flow. *European Two-Phase Flow Group Meeting, Ispra, Italy* (quoted by Hewitt 1982).

GROENEVELD, D.C. (1982). A general CHF prediction method for water suitable for reactor accident analysis. *CENG Report DRE/STT/ SETRE/82-2-E/DG.*

HETSRONI, G. (ed.) (1982). *Handbook of multiphase systems.* McGraw-Hill, New York.

HEWITT, G.F. (1978). Critical heat flux in flow boiling. *Sixth Int. Heat Transfer Conference*, Toronto **6** 143–71.

HEWITT, G. F. (1978). *Measurement of two-phase flow parameters.* Academic Press, London.

HEWITT, G.F. (1982). Flow regimes. In Hetsroni (1982).

HEWITT, G.F. (1982). Pressure drop and void fraction. In Hetsroni (1982).

HEWITT, G.F. and ROBERTS, D.N. (1969). Studies of two-phase flow patterns by simultaneous flash and X-ray photography. *AERE-M2159.*

HEWITT, G. F. and HALL TAYLOR, N. S. (1971). *Annular two-phase flow.*

HEWITT, G.F., LACEY, P.M.C., and NICHOLLS, B. (1965). Transitions in film flow in a vertical tube. *Proc. Smp. on Two-Phase Flow*, Exeter, UK.

HINZE, J.O. (1955). Fundamentals of the hydrodynamic mechanism of slitting in dispersion process. *AIChE J* **1** 289.

ISBIN, H.S., MOEN, R.H., WICKEY, R.O., MOSHER, D.R., and LARSON, H.C. (1958). Two-phase steam–water pressure drops. *Nucl. Sci. and Eng. Conf.*, Chicago.

IVEY, H.J. and MORRIS, D.J. (1962). On relevance of the vapour-liquid exchange mechanism for subcooled boiling heat transfer at high pressure. *AEEW–R137.*

KATTO, Y. and OHNE, H. (1984). An improved version of the generalized vertical tubes. *Int. J. Heat Mass Transfer* **27** 1641–8.

KUTATELADZE, S.S. (1948). On the transition to film boiling under natural convection. *Kotloturbostroenie* **3** 10.

KUTATELADZE, S.S.(1969). Boiling heat transfer. *Int. J. Heat Mass Transfer* **4** 31–45.

LIENHARD, J.H. (1981). *A heat transfer text book.* Prentice Hall, Englewood Cliffs, New Jersey.

LIENHARD, J.H. and DHIR, V.K. (1973). Extended hydrodynamic theory of the peak and minimum pool boiling heat fluxes. *NASA CR-2270.*

LIENHARD, J.H. and EICHHORN, R. (1976). Peak boiling heat flux of cylinder in a cross flow. *Int. J. Heat Mass Transfer* **1** 1135–41.

LOCKHART, R.W. and MARTINELLI, R.C. (1948-9). Proposed correlation of data for isothermal two-phase two-component flow in pipes. *Chem. Eng. Prog.* **45** 39–48.

MacBETH, R.V. (1963). Burnout analysis. III The low velocity burnout regime. *AEEW-R222.*

McKEE, H.R. and BELL, K.J. (1969). Forced convection boiling from a cylinder normal to the flow. *Chem. Eng. Prog. Symp. Ser.* **65** (No. 92) 222–30.

McQUILLAN, K.W. and WHALLEY, P.B. (1985). A comparison between flooding correlations and experimental flooding data, *Chem. Eng. Sci.* **40** 1425–40.

McQUILLAN, K.W., WHALLEY, P.B. and HEWITT, G.F. (1985). Flooding in vertical two-phase flow. *Int. J. Multiphase Flow* **11** 741–60.

Int. J. Multiphase Flow **10** 599–621.

MARTINELLI, R.C. and NELSON, D.B. (1948). Prediction of pressure drop during forced circulation boiling of water. *Trans. ASME* **70** 695–702.

MOALEM MARON, D. and DUKLER, A.E. (1984). Flooding and upward film flow in vertical tubes. II speculation on film flow mechanisms. *Int. J. Multiphase Flow* **10** 599–621.

MOSTINSKI, I.L. (1963). Calculation of heat transfer and critical heat flux in boiling liquids based on the law of corresponding states. *Teploenergetika* **10** (No. 4), 66–71.

NICKLIN, D.J. and DAVIDSON, J.F. (1962). The onset of instability in two-phase slug flow. *Instn of Mech. Engnrs Proc. Symp. on Two-Phase Flow*, paper 4.

NUKIYAMA, S. (1934). The maximum and minimum values of the heat transmitted from metal to boiling water at atmospheric pressure. Available as *Int. J. Heat Mass Transfer* (1966) **9** 1419–33.

OWEN, R.G. and LEE, W.C. (1983). Some recent developments in condensation theory. *Chem. Eng. Res. Des.* **61** 335–61.

PREMOLI, A., FRANCESCO, D., and PRINA, A. (1970). An empirical correlation for evaluating two-phase mixture density under adiabatic conditions. *European Two-Phase Flow Group Meeting, Milan* (quoted by Hewitt 1982).

PUSHKINA, O.L. and SOROKIN, Y.L. (1969). Breakdown of liquid film motion in vertical tubes. *Heat Transfer Soviet Research* **1** (no. 5), 56–64).

RADOVICH, N.A. and MOISSIS, R. (1962). The transition from two-phase bubble flow to slug flow. *MIT Report 7-7673-22*.

ROHSENOW, W.M. (1952). A method of correlating heat transfer data for surface boiling of liquids. *Trans. ASME* **74** 969.

ROHSENOW, W.M. (1956). Heat transfer and temperature distribution in laminar film condensation. *Trans. ASME* **78** 1645–8.

SCOTT, D.S. (1963). Properties of co-current gas-liquid flow. In *Advances in chemical engineering* **4** 199–277.

SILVER, L. (1947). Gas cooling with aqueous condensation. *Trans. Instn. Chem. Engnrs* **25** 30–42.

SMITH, S.L. (1971). Void fraction in two-phase flow: a correlation based upon an equal velocity head model. *Heat and Fluid Flow* **1** (No. 1) 22–39.

STARCZEWSKI, J (1965). Generalised design of evaporators—heat transfer to nucleate boiling liquids. *Brit. Chem. Eng.* **10** 523–31.

SUN, K.H. and LIENHARD, J.H. (1970). The peak pool boiling heat flux on horizontal cylinders. *Int. J. Heat Mass Transfer* **13** 1425–39.

TAITEL, Y. and DUKLER, A.E. (1976). A model for predicting flow regime transitions in horizontal and near horizontal gas–liquid flow. *AIChE J.* **22** 47–55.

THOM, J.R.S. (1964). Prediction of pressure drop during forced circulation boiling of water. *Int. J. Heat Mass Transfer* **7** 709–24.

TURNER, J.S. (1973). *Buoyancy effects in fluids.* Cambridge University Press.

WALLIS, G.B. (1961). Flooding velocities for air and water in vertical tubes. *AEEW-R123*.

WALLIS, G. B. (1969). *One-dimensional two-phase flow.* McGraw-Hill, New York.

WHALLEY, P. B. (1987). *Boiling, condensation and gas–liquid flow.* Oxford University Press.

WHALLEY, P.B. and McQUILLAN, K.W. (1985). Flooding in two-phase flow; the effect of tube length and artificial wave injection. *Physico-Chemical Hydrodynamics* **6** 3–21.

ZIVI, S.M. (1964). Estimation of steady-state steam void fraction by means of the principle of minimum entropy production. *Trans. ASME (J. Heat Transfer)* **86** 247–52.

ZUBER, N. (1959). Hydrodynamic aspects of boiling heat transfer. *AECU-4439*.

ZUBER, N. and FINDLAY, J.A. (1965). Average volumetric concentration in two-phase flow systems. *J. Heat Transfer* **87** 453–68.

ZUBER, N., STAUB, F.W., BIJWAARD, G., and KROEGER, P.G. (1967). Steady-state and transient void fraction in two-phase flow systems. *General Electric Report GEAP-5417*.

Index